GEOGRAPHICAL APPLICATIONS OF
AERIAL PHOTOGRAPHY

GEOGRAPHICAL
APPLICATIONS OF
AERIAL PHOTOGRAPHY

C. P. LO

CRANE, RUSSAK & COMPANY, INC.   New York

DAVID & CHARLES   Newton Abbot   London   Vancouver

To Seen Lim

© C. P. Lo 1976

First published 1976 by
David & Charles (Publishers) Limited
Brunel House   Newton Abbot
(ISBN 0 7153 7199)

Published in the United States of America by
Crane, Russak & Company, Inc
347 Madison Avenue
New York, New York 10017

(ISBN 0-8448-0872-5)
Library of Congress Catalog Card Number 75-37401

Printed in Great Britain
by A Wheaton Company, Exeter

Published in Canada
by Douglas David & Charles Limited
1875 Welch Street   North Vancouver   BC

CONTENTS

PREFACE

This book has been written for geographers who have an
interest in the application of aerial photography to
their problems. As the title indicates, the book gives
a survey of the various types of applications in
geography that are possible with aerial photography.
It also attempts to explain to geographers the underlying principles that have made the application possible.
The book also aims at introducing geographers to the
theories and methods of photogrammetry and photo-
interpretation which are essential for an evaluation of
the various applications. It is the author's belief
that before aerial photography can be employed intel-
ligently, the geographer should realise right from the
start the potentiality and limitations of the tool. The
photogrammetrist is always aiming for a higher degree of
accuracy that may be achieved through a certain
technique. The geographer should similarly get into
that spirit if he at all wishes to use the technique in
his problems. It is perhaps more appropriate now than
before to talk about photogrammetry, because geographers,
after the so-called quantitative revolution, are becoming more critical on data accuracies.

Although this book deals only with applications using
the photographic image obtained from an airborne platform
within the atmosphere, it should be pointed out that
non-photographic remote sensing, such as imageries
obtained from the thermal infra-red linescanner or the
sideway-looking radar system, as well as extra-
terrestrial photography obtained from satellites and
rockets in outer space have been growing in importance
in recent years. Together, they have exercised a great
influence on the trends and developments of photogrammetry and photo-interpretation. Their applications
in geography are even more varied and could only be
fully evaluated in another book.

7

In the course of writing the present book, the author has been intrigued by the vastness of the field and has realised that it is virtually impossible to cover all the different types of application within the limited scope of this book. It is hoped, however, that this book can indicate the right direction for the application of aerial photography to geography and stimulate discussions on the suitability of most of the methods.

C.P. Lo   February 1975

I GEOGRAPHY AND AERIAL PHOTOGRAPHY

AERIAL PHOTOGRAPHY AS A TOOL IN MODERN GEOGRAPHY

The 'Quantitative Revolution' in the mid-1960s has brought a 'New Geography' which takes the search for models and theories as a recurrent theme of study. This trend towards a nomothetic approach has enhanced the importance of aerial photography as a tool in collecting accurate and up-to-date data for testing geographical hypotheses and models.

The term *aerial photography* by itself is usually applied in a rather restricted sense to mean the taking of photographs from an airborne platform. In this book, its use is much wider to cover the entire field of the study of aerial photographs, and is therefore more or less synonymous with the term *aerial surveying*, or *air survey*, which embraces both the methods of photogrammetry and photo-interpretation.

The development of aerial photography has taken place comparatively recent, and has been heavily dependent on advances in optics and aircraft technology. When the first lens was ground in 1812, the era of photography began which ultimately led to the development of better lenses such as the distortion-free Hypergon lens covering a view of 72° and the invention of the Daguerreotype camera as a practical means of photography in the 1840s. In 1858 the first aerial photographs were produced in France by Nadar (Gaspard Felix Tournachon) using the Daguerreotype camera and a balloon. From these photographs the Frenchman Aimé Laussedat, generally known as the Father of Photogrammetry, applied the method of perspective to produce a map on paper.[1] Unfortunately, because the limitations of balloons or kites as camera platforms prevented the taking of enough photographs to cover all the area required, no further advances in mapping using aerial photography occurred until 1903 when the Wright brothers successfully developed the aeroplane and made flying possible. Until then, interest centred on terrestrial photogrammetry, ie the use of ground photographs for mapping. In 1909 the first photographs taken from an aeroplane were made. During World War I aerial photography rapidly expanded as its military importance was realised, which resulted in great improvements in the techniques of photo-interpretation. As aerial photography became more generally available, the advantage of aerial photography in giving a bird's-eye view, or the holistic view of our terrestrial environment became apparent. It immediately attracted the attention of earth scientists, and as early as 1920 the potentialities of aerial photography for use in archaeology, botany, geography and geology were made known. A good example of early geographical applications reflecting a typical 'regional approach' was the work by Lee, who used aerial photographs to describe the 'face of the earth'.[2]

Subsequent developments in aerial photography have always been aiming at improving the accuracy of the quantitative (or metric) as well as the qualitative (or descriptive) data extracted from aerial photographs. These have been achieved in three directions: (1) Camera lens design, (2) Photographic materials, and (3) Photogrammetric plotting machines.

1 *Camera Lens Design.* The angular coverage of the lens was increased, for example, the Topogon lens manufactured by Zeiss of Germany in 1933, covering an angle of 90°, and the Metrogon lens by Bausch and Lomb in the United States after World War II. This made possible wide-angle photography, which gives rise to a more favourable base/height ratio (ie the ratio of the separation between two successive exposure stations from the camera to the object) for better heighting accuracy than the narrow-angle (45°) photography. Another advantage is the considerable saving in the number of control points required in photogrammetric plotting since a smaller number of photographs is required to cover a

larger area. Today, the use of wide-angle photography
is generally accepted as the standard practice in
photogrammetry. In 1959 the Swiss firm of Wild manufactured the Super-Aviogon lens which covers a view of 120°,
thus increasing even further the angular coverage of
the camera (ie super-wide-angle camera).[3]

Meanwhile, the metric quality or the precision of the
lens has also been considerably improved through more
accurate computation made possible by the electronic
computer. Modern camera lenses are claimed to be
distortion-free; the Pleogon lens of Zeiss (Oberkochen,
West Germany) and Aviogon lens of Wild, for example,
claimed a distortion of between ±5 and 6μm.

2 *Photographic Materials*. The photographic process
makes use of the fact that visible light (which may be
considered to lie between 400 and 700nm of the
electromagnetic spectrum) can react with a silver halide
salt to form an invisible latent image, which, after
development by chemical means to reduce the silver
halide grains to silver, results in a negative image
with densities proportional to the intensities of light
received. The sensitivity to light of this chemical
emulsion is important. The speed of the emulsion today
has been increased by over 4 million times as compared
with 1839; for example, the Eastman Kodak Tri-X film
possesses a speed of ASA 200 (or DIN 25°).[4] The ability
of the emulsion to detect fine details (ie the resolving
power) has been greatly improved. It is also possible
to produce emulsions of different spectral sensitivity
(ie to respond to light of different wavelengths) for
different purposes of application, hence, the three
different types of sensitising known as *orthochromatic*,
panchromatic and *infra-red*. Even the recording of
colour is possible by combining in a film three layers of
emulsions which are separately sensitive to red, green
and blue light. Another great improvement, made notably
by the Eastman Kodak Company, is the dimensional stability of the film base on which emulsion could be coated
so that now photogrammetry can be released from using the
heavy glass plate needed in the past. Such a film base,
which is made of polyester material, is variously known
as Mylar, Estar, Cronar, etc, but is generally called
a Topographic Base.

11

3 *Photogrammetric Plotting Machines.* The rapid
production of topographic maps from aerial photographs
has been made possible by using machines which can solve
by analytical or analogous means the geometric problems
of aerial photographs in order to obtain correct
topographic maps, eg the Stereocomparator invented by
C. Pulfrich of Zeiss in Germany in 1900 and the
Stereoautograph by E. von Orel in Austria in 1908. The
most important development was *stereophotogrammetry*
which requires the use of stereopairs of photographs
for plotting in the machines (hence the name
stereoplotter). These photographs can be aerial or
terrestrial, and in 1923 a universal machine called the
Stereoplanigraph designed by Professor Bauersfeld was
successfully produced in Zeiss, Germany, which accepted
both terrestrial and aerial photographs in stereopairs
in plotting maps. This was followed by the emergence
of another machine - the Multiplex by Zeiss in 1933 -
which had a great impact on photogrammetric mapping in
the United States. Today, the variety of plotting
machines is even greater, notably the series of machines
manufactured by the Swiss firm of Wild (eg A7, A8, A9,
A10, B8, B9). Most important is the availability of a
cheaper range of stereoplotters (known as topographic
plotters) which are capable of achieving a reasonable
degree of accuracy for more general applications, apart
from topographic mapping using small-scale or medium-
scale photography. Examples are the Wild B8 and the
Kern PG2.

It is clear from the above examination of the three
major lines of development that the aerial camera is in
fact a powerful and precise surveying instrument which
is capable of recording the minute detail of our
terrestrial environment subject to the limitations of
the scale of photography and the resolving power of the
film. Thus, with the geographers' concern for the
terrestrial environment, the use of aerial photography in
geographical studies was well established in the period
just before World War II and the term *photo-geography*
was generally used to refer to the subject.[5] Not
surprisingly, interest mainly centred on regional
geography in which aerial photographs were used as a
kind of supplement to maps for the delineation of

regions. In modern terminology, these regions are areal classes which have uniform characteristics throughout, and the whole process of regionalisation is analogous to the technique of classification which is generally employed at an early stage by other natural sciences, notably botany and zoology (where the term *taxonomy* is used).[6] Therefore, the main use of aerial photographs at such a stage is to provide a base for regional inventories. Although the inventorying is necessarily descriptive, the approach is still useful today when a reappraisal of the man-land relationship and the maintenance of the ecological balance are called for in geography. Paradoxically, there is also the need to develop further the underdeveloped areas found within developed nations such as Canada (eg the Northern Lands) as well as in developing nations (so-called Third World countries). Natural resources inventorying and topographic mapping in these areas can benefit greatly from the appropriate use of aerial photographs. One of the modern approaches is to identify the repetitive patterns of similar tones and textures appearing on photographs over a broad area as *photomorphic units* (equivalent to geographic individuals), which can be employed for evaluation purposes.[7] The ability of aerial photographs to provide environmental data from the spectral, spatial and temporal standpoints is invaluable in a *new* regional approach - ie regional analysis in which a higher degree of objectivity and quantification is characteristic.[8]

THE IMPORTANCE OF STEREOMODELS

One of the most valuable properties of aerial photography is its ability to give rise to a stereomodel. The stereomodel is necessarily a mental model obtainable in one's mind by viewing a stereopair of photographs. It is made possible by the pair of human eyes which, by means of *convergence* (ie gazing at the object) and *accommodation* (ie focusing the eye-lens), fuse the two images together to form a three-dimensional model. Usually, the stereomodel is seen with the aid of a stereoscope to magnify and to maintain the three-dimensional impression. There are various theories explaining the formation of the stereomodel, and it is quite obvious that some physiological and psychological factors have been

13

involved when one perceives the three-dimensional model. Raasveldt attempted a reconstruction of the stereomodel in space by employing a geometrical theory of projection.[9] Modern research has shown that the image formed by the eye's optical system is sensed in the retina by a photo-chemical process as the first section of the visual pathway.[10] This information is transmitted along the visual pathway until it reaches the occipital cortex of the brain which permits the visual perception of the three-dimensional model. It should be noted that experience plays an important part in aiding stereoscopic or binocular vision. Thus the stereomodel is a spatial-analogue model giving a small-scale representation of the reality. Geometrically, the stereomodel as seen by an observer is incorrect because the photographs are usually placed on the flat table underneath the stereo-scope without taking into account any tilts that may have been present at the time of photography, and because the photographs are not generally viewed at a distance equal to the principal distance of the camera lens. An affine-deformed stereomodel (ie a linear scale deforma-tion of the model in the vertical direction) results, giving rise to different scales in planimetry and height. Another defect of such a stereomodel is that it is considered to be highly subjective. On the other hand, it is possible to rectify all these defects by using more sophisticated machines, such as the stereoplotter which can take out any tilts present and the densito-meter which can measure the tonal values from one part of the photograph to the next, thus making possible more objective interpretation.

THE DUAL FUNCTIONS OF AERIAL PHOTOGRAPHY IN GEOGRAPHY

This brief survey has revealed two major functions of aerial photography in geography today, namely, as a data-collecting tool and as an analytical tool. It is particularly noteworthy that aerial photography, by virtue of its special vantage-point, is an efficient surveying tool for inventorying and mapping. In other words, it is capable of collecting data from the spatial, temporal and spectral standpoints. On the other hand, the stereomodel as generated from special stereo-photography can be treated as an analogue model of the terrestrial environment, which can be continuously

14

simplified, by changing into successively smaller scales, to give rise to some conceptual models in geography, thus meeting the need for a more theoretical application of the 'New Geography'. As such, aerial photography is acting as an analytical tool in geography.[11] At the same time, aerial photography has been undergoing tremendous improvements, especially towards refinements in the techniques of photogrammetry and a more objective approach in photo-interpretation. All these have combined to enhance the value of aerial photography as a tool in modern geography.

II THE BASIS OF PHOTO-GRAMMETRIC TECHNIQUES

The potentialities of aerial photography as a tool in geography cannot be fully appreciated if the geographer fails to grasp the basic concepts of photogrammetry. It is not possible to give a full treatment of photogrammetry in this chapter, but attention is drawn to the more important topics.

By definition, photogrammetry is 'the science or art of obtaining reliable measurements by means of photography'.[1] This term first appeared in a German paper published in 1839 by Meydenbauer. The principles of photogrammetry are applicable to ground photographs (known as *terrestrial photogrammetry*) as well as aerial photographs (known as *aerial photogrammetry*). For geographical applications, as in topographic mapping, aerial photographs are usually preferred.

TYPES OF AERIAL PHOTOGRAPHY

There are basically two types of aerial photography: *vertical* and *oblique*. Vertical aerial photography is taken with the optical axis of the camera held in a vertical position. However, an unintentional and unavoidable inclination of the optical axis from the vertical usually occurs, thus producing a tilted photograph. Today, most commercial survey companies can reduce the tilt to less than 2° from the vertical. Oblique aerial photography is taken with the optical axis of the camera intentionally inclined to the vertical. There are two types of obliques, depending on how large the angular inclination is from the vertical. In *high-angle obliques*, the camera inclines at a much larger angle from the vertical. Thus, it points only slightly downwards and the photograph will normally contain the apparent horizon of the earth. In *low-angle obliques*, the camera inclines only at a small angle from the

16

vertical. Thus the camera points steeply downwards and
the photograph will not include the apparent horizon at
all (Fig 2.1). Despite this classification, all aerial

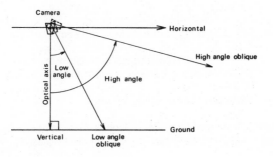

Fig 2.1 *Types of aerial photography*

photographs are perspective or central projections formed
as a result of the light rays passing through a fixed
point - the camera lens. This gives rise to scale
change throughout the resultant photograph. The oblique
photographs emphasise this perspective effect more
clearly than the vertical photographs, with distinctive
scale changes from the front to the back and from one
side to the other.

It is possible to combine vertical and oblique views
together in one shot by using two or more cameras in
synchronisation. Notable examples include the
Trimetrogon photography of the United States Air Force,
which utilised three Metrogon wide-angle cameras (f =
152mm) in one assembly with the central camera pointing
vertically downwards and the other two cameras pointing
to the left and right of the flight line at a depression
angle of about 30°;[2] the *nine-lens camera* of the US
Coast and Geodetic Survey with the central lens (f =
209.3mm) pointing vertically downwards to be surrounded
by the eight other lenses pointing at an angle of 38°
away from the vertical; and the split vertical

17

photography of the British Air Force which made use of
one or more pairs of cameras fitted on the aircraft to
point slightly outwards from the vertical.[3] These
multiple camera installations have been designed
specifically for military reconnaissance purposes and
are therefore not particularly favoured in mapping
because of the more complicated geometry involved.
However, geographers may find these systems useful for
their wide coverage and for macro-scale applications
where only qualitative data are required.

FACTORS CONTROLLING THE IMAGE
QUALITY OF AERIAL PHOTOGRAPHS

The Aerial Camera (Plate 1)

The aerial photograph is the product of a precision
surveying instrument - the aerial camera (Fig 2.2). As

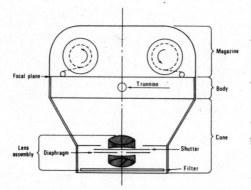

Fig 2.2 *Components of an aerial camera*
(after Moffitt, 1967)

the exposure has to be made whilst the aeroplane is in
motion, the aerial camera requires a fast lens, a quick,
efficient and dependable shutter and a high-speed
emulsion for the film. Hence, the image quality depends
on (1) the lens of the camera system, (2) the film type
used and (3) the development and printing processes.

1 *The Lens System*. The function of the lens is to
gather a selected *bundle** of light rays for each of an
infinite number of points on the terrain and to bring
each bundle to focus as a point on the focal plane
(Fig 2.3). In the aerial camera, the focus is always

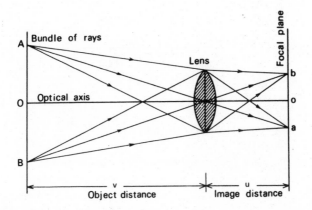

Fig 2.3 *Function of a lens in focusing
a point as an image*

set for an object distance of infinity (ie a fixed-focus
camera). It should be noted, however, that the actual
lens system is rather complex and that Fig 2.3 gives
only a simplified representation. The lens system in
fact contains two nodal points on the optical axis:
the front nodal point (N_1) on the object-space side and
the rear nodal point (N_2) on the image-space side
(Fig 2.4). Some actual examples of commonly used lens
systems are shown in Fig 2.5.

---

*A *bundle of light rays* is a collection of rays spreading
out in three dimensions (ie in different planes) like the
ribs of an umbrella, and should be differentiated from a
pencil of rays which is a number of rays radiating from
one point in two dimensions, ie in one plane like the
spokes of a wheel.

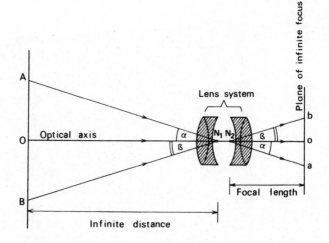

Fig 2.4 *Aerial camera lens system showing nodal points*

The amount of light which is admitted through the lens is controlled by the diaphragm and shutter. The *diaphragm* is a physical opening the size of which is adjustable by rotating a series of leaves. When the diaphragm is wide open, the lens resolution is greatest. The *shutter*, which is located between the lens elements and close to the plane of the diaphragm, controls the interval of time during which light is allowed to pass through the lens (ie exposure time). The *intra-lens shutter* is usually preferred for the aerial camera because it admits light to all parts of the negative instantly upon opening and cuts off light from all parts of the negative instantly at the end of the pre-set exposure time interval. Thus, it gives rise to a central projection which helps to preserve in the negative the precise relationship of all object points photographed.

The lens design involves lengthy computations, after which the lens has to be ground to the exact specifications and then the various elements assembled together. Human errors inevitably occur in the manufacture of the

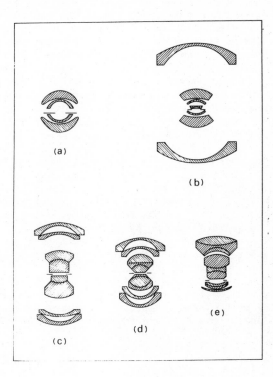

Fig 2.5 *Sections of some commonly used lens systems: (a) Bausch and Lomb Metrogon, f/6.3; (b) Zeiss Pleogon, f/5.6; (c) Wild Aviogon, f/5.6; (d) Wild Super-Aviogon, f/5.6; and (e) Wild Aviotar, f/4*

lens, for example, in grinding and polishing the surfaces of the optical glass and in maintaining a high consistency in its refractive index. For a wide-angle camera with 152mm focal length, the effects caused by differences in the refractive index could result in a distortion of 3-4μm in the image plane. Also, in assembling the lens elements together, the accuracy of the mechanical

21

alignment is influenced by the accuracy of machining the
necessary metal parts and by the fact that the mechanical
assembly must leave some space to allow for differential
expansion of glass and metal. All these errors cause
residual distortions which result in the lens not being
able to bring an image to focus on its theoretically
correct position in the focal plane. Most of these
distortions occur radially from the centre of the focal
plane (radial distortions), but there are also distortions which occur in a direction perpendicular to the
radial line (the tangential radiation), apparently caused
by faulty centring of the lens assembly.

Apart from the lens assembly which is a source of distortions in the resultant image, the film flattening
device and the film transport inside the magazine can
give rise to film deformations. It is essential that
the film be kept flat on the focal plane during exposure.
This may be achieved either by a pneumatic method using
over-pressure by pumping air out from the back of the
film to create a vacuum or by mechanical flattening by
means of pressure being applied on the back of the film
over a glass plate on the focal plane. But the film
flattening device is never perfect. In the pneumatic
flattening method, for example, the contact between the
back surface of the film and the reference plate may be
prevented or disturbed by improper functioning of vacuum
pumps or by particles trapped between the two surfaces.
When this is the case, image shift occurs, ie a point to
be imaged in $\underline{A}$ is actually imaged at $\underline{A}'$ and, if parallel
projection is used in the mensuration process, appears
to be located at $\underline{A}''$ (Fig 2.6a); a map of the displacement
vectors over the photograph reveals that this occurs
radially, and large localised errors will show up clearly
at those parts where improper flattening occurs (Fig
2.6b).[4] As for the film transport in the magazine, an
appreciable force is continually applied to pull the film
across the focal plane, thus leading to elastic extension
of the film prior to the exposure and to contraction
afterwards.

Because of the presence of these sources of distortion,
the aerial camera has to be calibrated before being used
to obtain aerial photographs for photogrammetric purposes.
This calibration makes use of elaborate instruments under

22

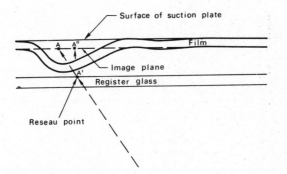

Fig 2.6a *Image shift resulting from improper film flattening in an aerial camera (A → A") (after Ziemann, 1972)*

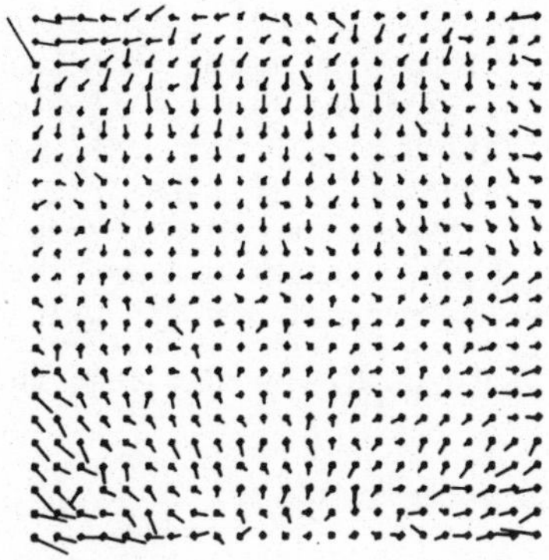

Fig 2.6b *Deformation pattern of an aerial photograph including two large local errors resulting from improper flattening of the film in the aerial camera (after Ziemann, 1972)*

controlled laboratory conditions to determine accurately the principal point or the collimation centre on the focal plane, the focal length and the projection centre of the lens system. Thus one establishes the mathematical relationship between points on the image surface of a *real* camera and those of an *ideal* camera, so that appropriate corrections can be made to their positions.

2 *The Film Type*. The film type used also plays an important part in determining the quality of the resultant image. The photographic emulsion is silver halide salt suspended in gelatin and is coated on some suitable supporting base material. When light, a form of radiant energy of specific wavelengths in the electromagnetic spectrum, falls on this, a chemical reaction occurs which converts the emulsion into a latent image. This latent image is composed of a small aggregation of silver atoms formed on the surface or the interior of the grains. They are the 'sensitivity specks' acting as development centres. In developing, some chemical reducing agents are used so that the small black sensitivity specks of silver are precipitated.[5] Since unexposed areas are not reduced to silver at all, the reaction is totally proportional to the amount of light falling on the emulsion. The result of the development is a negative image, from which a positive point can be obtained. The variations in the intensities of the light on the sensitised materials give rise to different degrees of darkness, the tones of which can be measured and expressed by a number called the *density*. The density is defined in terms of *opacity* which measures the proportion of light passing through the film. Thus, density can be expressed mathematically as

$$D = \log_{10} O = \log_{10} \left(\frac{I_o}{I_t}\right)$$

where D is the density, O is the opacity, I_o is the incident light and I_t is the transmitted light. The higher the density, the darker the film.

Closely related to density is the concept of *contrast* which measures the actual difference in density between the high-lights and low-lights or shadows on a negative

or photographic print.[6] This can be studied together
with the *speed* of the photographic material by means of
the *characteristic curve*, which portrays graphically the
relationship between density and the time of exposure
(measured on a logarithmic scale)(Fig 2.7). The result

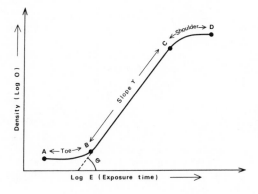

Fig 2.7 *A characteristic curve*

is an elongated S-shaped curve which shows three distinct
portions as (a) the toe, (b) the slope or Gamma (γ) and
(c) the shoulder. The slope portion which is more or
less a straight line can be extended to meet the
abscissa to form an angle θ. This is generally known as
Gamma (γ) which is in fact the tangent of angle θ. The
steeper the slope (ie the higher the Gamma), the greater
the contrast. Thus a fast film (with hard gradation,
short exposure time and large density range) has a bigger
Gamma than a slow film (with soft gradation, long
exposure time and small density range).

These observations lead to other considerations of the
properties of the sensitised materials used in aerial
photography such as the speed, spectral sensitivity and
the resolving power. The *speed* of the emulsion may be
thought of as a measure of its sensitivity, and as yet
there are no satisfactory means of measuring the speed of
the films used for aerial photography, since the ASA
(American Standards Association) or DIN (German Institute)

25

methods of measuring film speed do not relate to the
special conditions under which the film is used. The
Eastman Kodak Company has developed an Aerial Exposure
Index in which the speed point is based on the point on
the characteristic curve where the slope is 0.6 of the
value of Gamma.[7]

The emulsion used is not sensitive to all parts of the
visible light spectrum, and its response to light of
different wavelengths depends on its *spectral sensitivity*.
All silver halide photographic materials have inherent
sensitivity only to the short-wavelength blue light.
For lights of longer wavelengths (such as green, yellow
and red), sensitising dyes have to be added before they
can be sensed. Thus, in the class of monochrome films,
there are three types of sensitising: (a) *orthochromatic*,
(b) *panchromatic* and (c) *infra-red*. The orthochromatic
will respond to blue and green lights (wavelengths from
about 400 to 600nm); the panchromatic will respond to
blue, green and red lights (from below 400 to 700nm);
the infra-red will respond to all lights in the visible
spectrum as well as the invisible near infra-red (from
below 400 to about 900nm) (Fig 2.8). Each of these
types of film is useful in aerial photography, but for
more general purposes the panchromatic film is preferred.
There is still another class of film - the *colour film* -
which can depict the natural colours of the landscape.
It is a more complex type of film than the monochrome,
being made up of three layers of emulsion sensitised
separately to blue, green and red lights. Its use in
aerial photography is relatively recent and still requires further investigation. Obviously, the types of
film used can significantly affect the quality of the
resulting image.

To evaluate the quality of the image it is usually
necessary to consider the *resolving power* of the lens as
well as the film. The resolving power of a film is a
subjective measure of image 'sharpness', expressed as the
number of line pairs per millimetre which can be distinguished in the image. This requires the use of a test
pattern which usually consists of three thick parallel
lines on a contrasting background. The width of the
lines is equal to the space between them. A basic
pattern of vertical lines is placed next to a pattern of

26

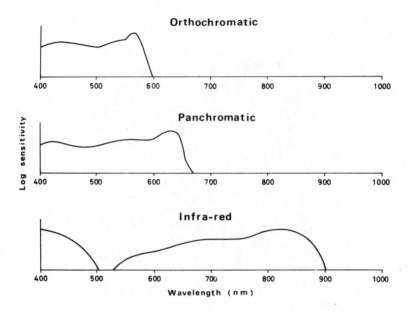

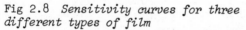

Fig 2.8 *Sensitivity curves for three different types of film*

horizontal lines, and the combination is repeated in a range of sizes over the target area, as shown by the US Geological Survey example which has been employed for camera calibration (Fig 2.9). For aerial photography, low-contrast resolution is more significant and a ratio of line-to-space luminance of 1.6 to 1 is used. But this subjective measure is not entirely satisfactory as it is dependent on the *contrast* between the lines and the background and does not therefore reliably indicate visual image sharpness.[8] Thus, an alternative method of evaluating image quality - the *Modulation Transfer Function* (MTF) - is now preferred.[9] The MTF or sine-wave response is a curve indicating the degree to which image contrast is reduced as spatial frequency is increased.

The MTF concept makes use of the fact that the brightness variation across a line chart results in a series of square-waves (Fig 2.10a). From the Fourier Theorem,

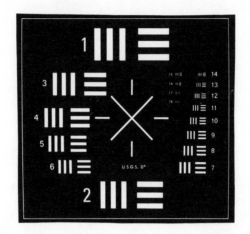

Fig 2.9 *A resolving-power test pattern for camera calibration employed by the United States Geological Survey. Numerals indicate the number of line pairs per millimetre*

these square waves can be synthesised from a series of sine waves (Fig 2.10b) and they correspond to sine waves of a fundamental frequency, together with their associated harmonics. It should be noted that the photographic image of a single narrow band of light near the limit of resolution will show a density such as that shown in the unbroken line in Fig 2.10c in which the ideal image is shown dotted. Thus, in determining the modulation transfer, a sinusoidal target is used which consists of a series of density variations between a constant maximum and a constant minimum, with the high-density peaks coming closer together as we move across the density variations (Fig 2.11a). The density profile in each element of the test target is sinusoidal - hence the term sine-wave response. If the film or lens under testing is perfect, all the peaks will be resolved. But this is not the case and at some stage the film or lens will fail to resolve the peaks of density. The density variations in the image of the test become as shown in Fig 2.11b. As

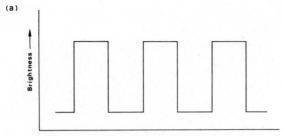

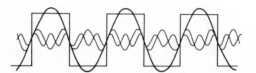

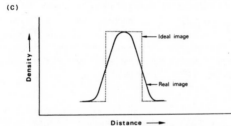

Fig 2.10 *(a) Brightness variation graph:
a square wave form (b) Breakdown of line
test-object into sine-curve components
(c) Density distribution of the photographic image of a single narrow band of
light near the limit of resolution: real
and ideal images compared (after Mullins,
1965)*

the peaks get closer, the ability of the film or lens to
resolve the peaks of density becomes less and less. A
photometer called the microdensitometer is required to
measure the density difference between peaks and troughs
at each line frequency. The values of maximum density

29

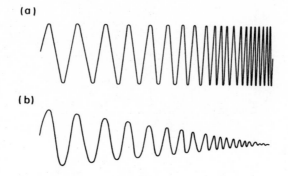

(a)

(b)

Fig 2.11   *(a) Ideal sinusoidal wave form*
(b) Sinusoidal wave form in reality

(D max) and minimum density (D min) at each frequency,
which can be determined directly from the microdensitometer trace, are converted to linear relative exposures
by means of the characteristic curve (ie density against
log E) to give maximum and minimum relative exposure
values (I max and I min) for the following formula of
modulation transfer (MT) factor:

$$\text{MT factor} = \frac{\text{I max} - \text{I min}}{\text{I max} + \text{I min}} \times \begin{array}{l}\text{(normalisation} \\ \text{factor to give} \\ \text{100\% modulation} \\ \text{at zero frequency)}\end{array}$$

The normalised MT factors are plotted against spatial
frequency to produce the sine-wave response curve
(Fig 2.12). Thus the curve indicates the response at all
frequencies, whereas the resolving power only indicates
the highest frequency visible.

   The major merit of the MTF lies in the possibility of
its being employed to evaluate the whole photographic
system, because individual MTF curves can be produced for
lens, films, image motion, and other relevant variables
in the camera system. A total system MTF is derived by

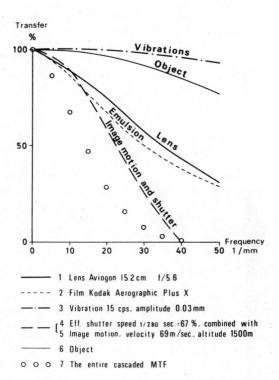

———————— 1 Lens Aviogon 15.2 cm f/5.6

- - - - - 2 Film Kodak Aerographic Plus X

—— · —— 3 Vibration 15 cps. amplitude 0.03 mm

— — [4 Eff. shutter speed 1/280 sec.=67 %, combined with
     [5 Image motion. velocity 69 m/sec., altitude 1500 m

———————— 6 Object

o o o 7 The entire cascaded MTF

Fig 2.12 *The MTF components of an aerial
photograph (after De Belder et al, 1972)*

multiplying the responses of the appropriate lens, film,
and other curves frequency by frequency in a process
known as *cascading*.[10] The result is a single MTF for
that combination of lens and film, etc. An actual
example is shown in Fig 2.12 where five components of an
aerial photograph are cascaded.

3 *The Development and Printing Process.* The development process detects and amplifies the latent image of
the exposed film to produce a visible silver image, which
is in negative form. The procedures involve wetting and
drying, both of which can have harmful effects on the

overall stability of the resulting negative. At a constant relative humidity, film dimensions increase with an increase in temperature, and decrease with a drop in temperature. Since the film is an elastic material coated with a gelatin layer, this overlying layer of emulsion exerts a compressive force on to the support at lower relative humidity (thus causing the emulsion to dry up quickly), which results in irregular (or non-linear) changes. In developing, the structure of the emulsion is modified by wetting, removal of unexposed silver halide and drying, thus resulting in a change of thickness and modulus of the emulsion layer, which reduces the compressing forces exerted by the emulsion. Also the drying needs to be carefully controlled because overdrying or forced drying results in lower relative humidity and tends to expand the film, whilst slowly dried film will usually show shrinkage. Today films are usually considered stable with the supporting base being made of polyester material, such as cellulose acetate butyrate in Kodak's Estar-base films whose non-uniform dimensional change should not exceed 3μm in a 240mm frame on average and is less than 8μm at any point in maximum. This is generally known as the *Topo Base Film*.[11] But even the Estar-base film is intentionally stretched in manufacture in both length and width to avoid a systematic orientation of its molecules. Further deformation occurs as aerial films are processed in the roller-transport processing machine, causing a lengthwise tension whilst wet and during drying.

From the negative it is possible to obtain a positive print on glass (known as a *diapositive*) or on paper. It can be produced in printers which project the negative through a lens or in contact printers. Projection printing gives rise to distortion due to displacements. Generally, contact printers with centrally projected light are used. These often allow automatic dodging of the photographs, a procedure which reduces highlights and shadows in the negative in order to improve the readability and interpretability of the printed diapositive by bringing out finer details.[12] Contact printing with automatic dodging can easily give rise to errors due to, for example, the film not being made to lie completely flat on the printer stage. On average, the magnitude of the error is from approximately ±2 to

Plate 1 The Wild RC-10 camera in operation *(Wild Heerbrugg Ltd)*

Plate 2 An example of the slotted template laydown

Plate 3 LUZ Aero Sketchmaster *(Carl Zeiss, Oberkochen)*

Plate 4 Zoom Transfer Scope *(Bausch and Lomb)*

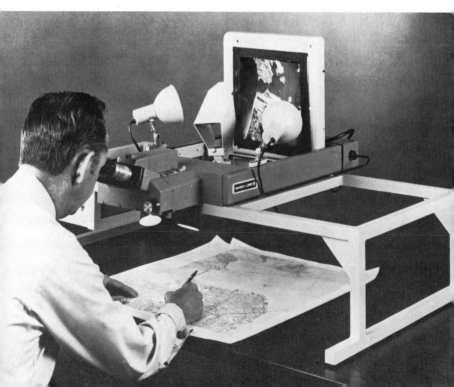

4µm. Experiments carried out by Corten at ITC in the Netherlands have shown that after positive contact copying (on paper or on glass), degradation in image detail occurs.[13] It has been suggested by Ziemann that printing with parallel light (instead of centrally projected light) can reduce the printing error by almost 50 per cent.[14]

Flying for Aerial Photography

Apart from the combined effects of the lens, the film and the development process on the quality of the resulting photographs, there are still other considerations, the major one being the manner in which the aerial photographs were taken. The task of flying for aerial photography is very demanding. It is necessary to satisfy the following conditions: (1) the aeroplane has to fly in a straight line, and at a constant speed so that a forward 60 per cent overlap of the photographs is maintained; (2) it has to turn round at a predetermined distance and to maintain the second flight line parallel to the first and to have a side-lap of 20 per cent; (3) the flying height is predetermined and should be constant throughout; and (4) the aerial camera should not be tilted at the time of exposure (Fig 2.13). To

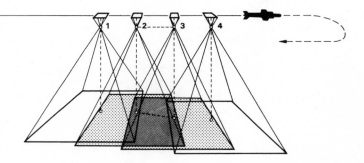

Fig 2.13 *Flying for air coverage*

achieve all these conditions, very skilful navigation of

the aeroplane is required. Modern airborne navigational equipment - the Doppler and Inertial systems - has to be used. In fact it very often happens that, owing to the presence of air currents, drifting of the aeroplane will occur, causing it to fly off course, and gaps in aerial coverage will occur (Fig 2.14 a,b). This defect can be made up only by turning the aeroplane round to compensate for the drift. To prevent crabbing from occurring, the aerial camera should also be rotated on its mount to allow for this angular change of the aeroplane (Fig 2.14c). It is important to note that the exposure is made as the aeroplane is moving, and that *image movement* will occur. This may or may not be detected according to the photographic scale, since the human eye cannot see any displacements less than 0.05mm. In general, for large-scale photography, image movement will become more prominent.[15] On the whole, for the most efficient execution of survey flights, Heimes suggests that the calibration of the aircraft in flight under the anticipated operational conditions should be carried out beforehand.[16]

Another major consideration is the *atmospheric effects* as the atmosphere is an extremely complex medium through which light rays pass. One of these effects is *photo-grammetric refraction* caused by the refraction of light rays away from the vertical when passing through air of decreasing density before reaching the aerial camera. This results in a small angle between the incoming refracted ray and the straight line connecting the ground point and the camera station. This angle is thus a function of the refractive index of the air in all points along the path of a ray. This can be eliminated only by a scale change, introduced, for example, by changing the focal length of the lens. This error can be avoided if photography is carried out only in relatively stable atmospheric conditions.

Another effect is *scattering* which involves deflection or absorption and re-emission of the light by gas molecules and dust particles in the atmosphere. As these particles and molecules are small with a size less than 0.1λ (λ is the wavelength of light), Rayleigh scattering occurs which is indicative of clear atmospheric conditions characterised by a blue sky as a result of

34

a
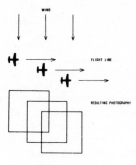

WIND

FLIGHT LINE

RESULTING PHOTOGRAPHY

b

INTENDED FLIGHT LINE

FLIGHT LINE

c
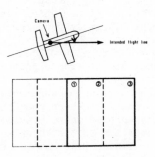

Camera

Intended flight line

① ② ③

Fig 2.14   (a) Drifting (b) Crabbing
(c) Correction by rotating the camera
mount

the blue light being scattered more. As the particles get larger in size, the scattering is more pronounced. Another type of scattering called Mie scattering is produced by particles with a diameter between 0.1λ and 25λ, ie approaching or exceeding the wavelength of the light, which include atmospheric aerosols, dust, haze, smoke, etc. A sky that appears white to red results. Mie scattering generally occurs in the lower atmosphere (below 4,560m) whilst Rayleigh scattering is generally found in the higher atmosphere (about 9,120m). These different types of scattering can adversely affect the quality of the photographic image obtained with an aerial camera by reducing the contrast. This undesirable effect can be eliminated by placing a minus-blue filter in front of the lens. This minus-blue filter is in fact a yellow filter made of high-quality coloured glass which can absorb the short-wavelength violet and blue lights. There are also other kinds of filter such as the ultra-violet filter to take out lights of short wavelength and the red filter which cuts out all lights except the red. For colour aerial photography, colour filters are used to raise or lower the colour temperature of the light coming into the lens to that of the standard colour temperature. As an example, a filter of the blue colour system will raise the temperature, whilst a filter of the amber system will lower the temperature. Thus, in the US Coast and Geodetic Survey, a filter which can cut out all lights below about 0.42μm is used for about 90 per cent of aerial colour photography, and a 'peach-shaded' filter which can cut out light below 0.38μm is employed for early-morning and late-afternoon photography.[17]

Today there are auxiliary instruments used in conjunction with the aerial camera when photographs are being taken. The purpose is to provide supplementary data on the orientation of the aeroplane at the moment of each exposure. These instruments are the *Statoscope*, the *Airborne Profile Recorder* (APR), the *Horizon Camera* and the *Gyroscope*. The Statoscope is a more sensitive device than the altimeter and is used to measure the variations of pressure relative to the first exposure. In this way, the variation in flying height of the aeroplane from Mean Sea Level in its course of flight is known. It is possible to determine the flying height up to an accuracy of ±1 to 2 metres. The Airborne Profile

Recorder makes use of radar which is transmitted from the
aeroplane to the ground and is then received back by the
aeroplane. The time taken by the radar signals in
returning is measured. Since the velocity of the radar
pulse is known, the distance of the aeroplane from the
ground surface can be computed. If a continuous
recording of these height readings is made during the
flight, a profile of the terrain can be drawn. By combining this with the statoscope record which indicates
the variations in the flying height of the aeroplane, an
accurate determination of the profile of the ground from
Mean Sea Level is possible. The Horizon Camera is used
in conjunction with the aerial camera to take photographs
of the horizon, from which the amount of tilt of the
aeroplane can be determined accurately. Finally, the
gyroscope is used to stabilise the aerial camera and to
ensure its verticality. Also, the deviation of the
camera from the vertical can be measured. All these
auxiliary instruments can help in the absolute orientation of the pair of aerial photographs when it is used in
plotting. But they add considerably to the cost of
aerial photography. For a more detailed discussion of
these auxiliary instruments, one is referred to the
papers by Schermerhorn,[18] Kennedy,[19] Eden,[20] and Trott.[21]

NON-CONVENTIONAL AERIAL PHOTOGRAPHY

The above discussion has been based on the use of an
aerial camera whose geometric characteristics are exactly
known from calibration and which is capable of producing
a single perspective (or central) projection in the
resulting photograph. This type of aerial camera is more
properly called the *photogrammetric camera* and is usually
employed for *conventional* aerial photography for topographic mapping. In contrast, there are also non-
conventional aerial cameras which are less suited for
topographic mapping by photogrammetric methods but may
be useful as a source of descriptive information for
geographical applications. Worth mentioning are the two
non-conventional photographic systems: (1) the Continuous
Strip Camera (or Sonne Camera) and (2) the Panoramic
Camera.

Continuous Strip Camera

This camera is a type of reconnaissance camera for military purposes which has been designed for low-altitude, high-speed photography. It exposes a continuous photograph of the terrain from the aircraft by passing the film over a stationary slit in the focal plane of the lens at the speed of the aircraft flight.[22] This compensates for the image motion relative to the terrain. There is no shutter in the camera but the aperture of the lens and the size of the slit can be varied. The slit is usually very small so that only a narrow 'ribbon' or 'strip' of the terrain is exposed on the film. As the aircraft moves forward, a long continuous photograph is produced by successive integration of these narrow ribbons, hence the name of the camera. It is obvious that with this type of camera the single-point perspective projection is destroyed as a result of the moving film. It is not possible to produce stereopairs of photographs using a single-lens cone for the camera. If stereo-photographs are required, a dual-lens cone with one lens displaced forwards and the other backwards has to be employed in order to create a stereobase in the direction of the flight. But because of the extended stereobase, the stereomodel cannot be obtained without a specially constructed stereoscope as an aid. It is particularly important to eliminate tilts at the time of photography, otherwise gaps will occur in the resultant coverage.

Panoramic Camera

The panoramic camera is another photographic system for military aerial reconnaissance. Basically, it consists of a moving lens and a slit which sweeps out a cylindrical path on to a film which may be stationary or moving at the same speed as the aircraft (ie with Image Motion Compensation) (Fig 2.15). The points on the ground are no longer imaged at their correct perspective positions on the film but become displaced according to the cylindrical shape of the negative film surface and the scanning action of the lens. Thus, in Fig 2.15, the ground point P is imaged as P'' with curved coordinate distances of Xp'' and Yp'' from the centre of the cylindrical film surface which may be compared with the

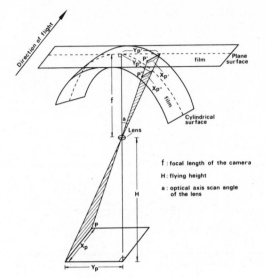

f : focal length of the camera

H : flying height

a : optical axis scan angle of the lens

Fig 2.15 *Basic geometry of panoramic photography*

corresponding perspective position and straight line
coordinate distances on a plane film surface (ie P', Xp'
and Yp'). This type of positional displacement is known
as *panoramic distortion* and an example of recording a
unit grid on *flat* ground (Fig 2.16a) with panoramic distortion is shown in Fig 2.16b. In addition, the forward
motion of the aircraft during the time of scanning also
modifies the position of points owing to panoramic distortion, thus giving rise to a *sweep positional distortion* which affects the Y-centre line as shown by the
dotted S-curves in Fig 2.16b. In the case where Image
Motion Compensation is incorporated, a third distortion
also occurs owing to the lateral movements of the lens or
focal plane (the film). This modifies the position of
points owing to both panoramic distortion and sweep
distortion in the X-direction but not in the Y-direction.
The overall effect of all these distortions in modifying
the unit grid on flat ground is shown in Fig 2.16c.[23]
It is obvious that for the panoramic photographic system,

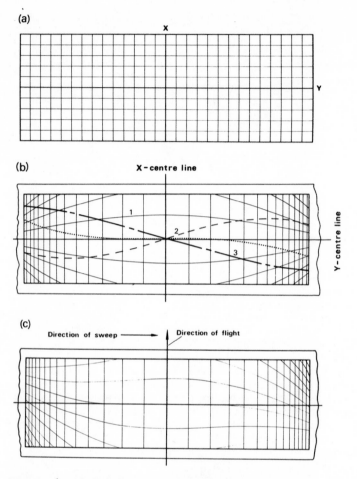

Fig 2.16 *Distortions for panoramic photography (after Hovey, 1965): (a) Original grid on the ground (b) Panoramic distortion effect (1), sweep positional distortion effect (2), and image motion compensation effect (3) (c) Pattern of the grid in (a) with all distortions combined*

the single-point perspective projection is not preserved and that the internal geometric characteristics of the system are difficult to determine exactly.

GEOMETRY OF THE CONVENTIONAL AERIAL PHOTOGRAPH

A conventional aerial photograph is generally regarded as a *central projection*, the properties of which provide the basis for the mathematical treatment of photogrammetry. But such a consideration is only an idealisation and is conditional upon the following assumptions:

1 The projection centre is in or near the lens. In reality, as we have seen earlier, no such centre exists, and the bundle of rays passes through different parts of the lens as shown in Fig 2.3.

2 A pencil of rays is represented geometrically as only a straight line as shown in Fig 2.4. In reality these rays are deviated in some way in the lens as a result of lens distortion.

3 A point in the object space is imaged by a plane in the negative space. In fact, there is no image plane but an image zone with a definite thickness due to the emulsion; and a collection of points rather than one point is imaged.

4 The emulsion base (either film or glass plate) is taken to be a perfectly flat surface and highly stable. In fact, deformations occur.

Scale

By making these assumptions, the aerial photograph is a graphic record of the light rays or, in mathematical terms, a presentation of the relationship between the *aerial camera* and the *ground*. Thus for a truly vertical photograph (Fig 2.17), the geometry of the relationship can be expressed as

$$\frac{pt}{TP} = \frac{op}{oP} = \frac{a}{b} = \frac{\text{image distance}}{\text{object distance}} \qquad (2.1)$$

In order that a sharp image can be obtained, Newton's Lens Equation has to be satisfied, ie:

$$\frac{1}{a} + \frac{1}{b} = \frac{1}{f} \qquad (2.2)$$

where f is the focal length of the camera. As the air camera has a fixed focus for the object distance at

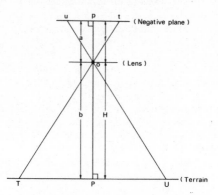

Fig 2.17 *Diagrammatic representation
of the relation between the photograph
negative and the terrain*

infinity, the value for b becomes infinity, thus:

$$\frac{1}{a} + \frac{1}{\infty} = \frac{1}{f}, \text{ ie } a = f,$$

which means that the image distance is equal to the focal
length when the object distance is at infinity. From
this the scale (S) of the aerial photograph can be found
to be, by substitution in 2.1:

$$S = \frac{f}{H} \qquad (2.3)$$

From Fig 2.17 it is noteworthy that the optical axis of
the camera (poP) is vertical to the terrain plane which
is considered to be flat (ie no relief). The perspective
centre of the lens which gives the geometric centre of
the aerial photograph is therefore point p. This is
generally known as the *principal point*, which can be
fixed on the photograph by joining the opposite pairs of
fiducial marks registered on the sides of the aerial
photograph by the aerial camera during exposure
(Fig 2.18). The principal point is normally taken as
the origin of a coordinate system (known as the photo or

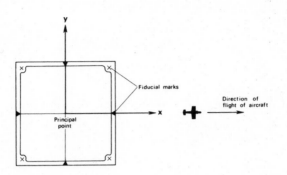

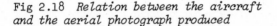

Fig 2.18 *Relation between the aircraft and the aerial photograph produced*

plate coordinate system, as distinct from the ground or terrestrial coordinate system). The X-axis of the photograph is the line between opposite fiducial marks which are parallel to the direction of the aircraft flight, whilst the Y-axis is that line normal to the X-axis (Fig 2.18). To take into account the other direction of movement of the aircraft, ie up and down, a third axis - the Z-axis - needs to be introduced. This also passes through the principal point and is perpendicular to the plane of the photograph. This coincides with the optical axis of the camera. Thus, any movement of the aircraft can be defined with reference to this three-axis coordinate system.

Tilts

It is obvious from the earlier discussion on flying for aerial photography that it is impossible to obtain tilt-free photographs. The movements of the aircraft with reference to the three mutually perpendicular X-, Y- and Z-axes give rise to the three major types of tilt (Fig 2.19):

1 φ*(Phi)-tilt or longitudinal tilt or tip* is the first or primary rotation taken about the Y-axis;

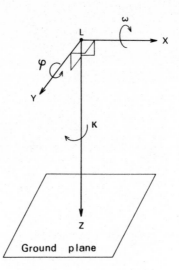

Fig 2.19 *Generalised coordinates of a
rigid body showing three types of
tilting about the three coordinate axes*

2 ω*(Omega)-tilt or lateral tilt* is the second or the
secondary rotation taken about the X-axis; and
3 κ*(Kappa) or swing* is the third or tertiary
rotation taken about the Z-axis.
It should be noted that the rotation angles shown in
Fig 2.19 are measured in a clockwise direction and
indicate positive direction, the so-called *right-handed
system* adopted for a negative.

With tilting, the geometrical relationship between the
camera and the ground will be changed as shown in
Fig 2.20 which should be compared with Fig 2.17. The
tilt is in fact the angle formed between the optical axis
and the *vertical* line (ie *t* in Fig 2.20), or the angle
which the plane of the photograph makes with the horizontal plane.

Clearly, with the presence of tilts, the optical axis
of the camera is no longer pointing vertically, and where

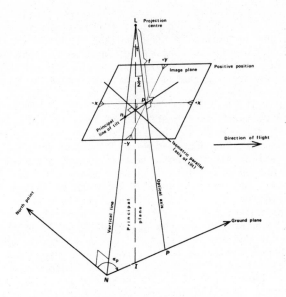

Fig 2.20 *Image-ground relationship*
for a tilted photograph

the vertical line cuts the plane of the photograph a
nadir point (*n*) is found. It is the point on the photograph vertically beneath the camera. A corresponding
ground point *N* also exists. The optical axis and the
vertical line together constitute a vertical plane (*nLp*
or *NLP*) known as the *principal plane*. The intersection
of the principal plane with the plane of the photograph
gives a line called the *principal line of tilt*. The
direction of the principal plane of the photograph on the
ground can be obtained if the north point is known and
its bearing is the clockwise angle measured horizontally
about the ground nadir point, *N*, as angle αp. This is in
fact the ground direction of the tilt since the angle of
tilt lies in the principal plane. The angle of tilt can
be bisected and the angle bisector meets the photograph
along the principal line at a point called the *isocentre*
(*i*) with a corresponding point on the ground at *I*. The
point *i* is at a distance of f.tan(*t*/2) from the principal
point where *f* is the focal length, with reference to the

right-angled triangle *Lnp* in Fig 2.20. From the isocentre a perpendicular line can be drawn which indicates
the *axis of tilt*. This horizontal line which passes
through the isocentre is known as the *isometric parallel*
because the tilting has no effect on its axis and hence
the scale along it is not changed at all. Other lines
that can be drawn parallel to this axis of tilt are known
as *plate parallels*.

The graphic and metric effects of the tilts on the
resulting photograph may best be seen by assuming that
the ground is flat and is covered by a square net. The
truly vertical aerial photograph will give a correct
representation of this at a reduced scale as shown in
Fig 2.21a. But if the aerial photograph is tilted, the
representation of the shape of the terrain will be dis-
torted as shown in Fig 2.21b. It is found that the scale
changes regularly along the principal line of tilt so
that at one end the scale is uniformly too small whilst
at the other the scale is uniformly too large. But on
the axis of tilt no scale changes occur (hence the name
isometric parallel). All displacements resulting are either inwards or outwards from the isocentre depending on
whether it is below or above the axis of tilt. With
reference to Fig 2.22, one can mathematically determine
the scale of the tilted photograph at the nadir point (n),
principal point (p), isocentre (i) and any point $(m$ or $q)$
on the plate parallel by the following formulae (which
can be easily verified by employing simple trigonometric
relationships):

(1) scale at the nadir point $= \dfrac{no}{oN}$

$$= \dfrac{f}{H.\cos t}$$

(2) scale at the principal point $= \dfrac{op}{Po}$

$$= \dfrac{f.\cos t}{H}$$

(3) scale at the isocentre $= \dfrac{io}{oI} = \dfrac{f}{H}$

a

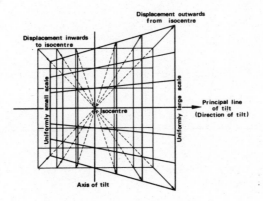

b

Fig 2.21 *(a) Correct representation of
the flat terrain by a truly vertical
aerial photograph (b) Tilted photograph
superimposed on the truly vertical
photograph*

and (4) scale at *m* on a plate parallel (ie on the downward side of the isocentre) $= \dfrac{mo}{oM}$

$$= \frac{f}{H} \, . \, \cos t \left(1 - \frac{r}{f} \, . \, \sin t\right)$$

47

scale at q on a plate parallel (ie on the upward side of the isocentre) $= \dfrac{qo}{oQ} = \dfrac{f}{H} \cdot \cos t (1 + \dfrac{r}{f} \sin t)$

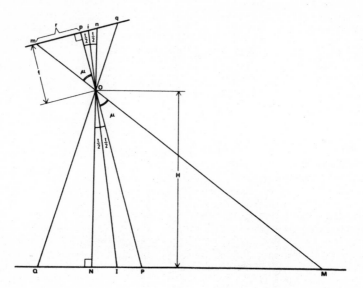

Fig 2.22 *Geometric analysis of a tilted photograph*

Influence of Relief

So far, our discussion of tilting has ignored the effects of relief. It is important to note that the map is essentially an *orthogonal projection* as shown in Fig 2.23 which is in contrast to the *central projection* of the aerial photograph. Assuming the case of a truly vertical aerial photograph, the introduction of the relief causes scale change and relief displacements as shown in Fig 2.24. The position of the hilltop at A' should be represented on the map as the orthogonal projection of A' at A on the datum level; but on the negative plane of the aerial photograph, it is imaged as a' as a result of the central projection through o. The correct position

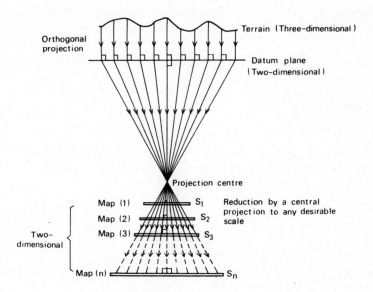

Fig 2.23 *Orthogonal projection method of map making*

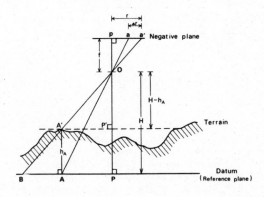

Fig 2.24 *Effect of relief displacement*

for the hilltop A on the negative plane should be a. The distance aa' on the negative plane is called *relief displacement* (Δr). When the relief is introduced, the scale of the photograph is defined by the ratio $\frac{a'p}{A'P'}$, which by similar triangles $a'op$ and $A'P'o$, is equal to

$$\frac{f}{H - h_A}$$

This only gives the scale at one elevation. For a large area with changes in elevation, the average scale is required, which can be found by averaging the heights of the various terrain points, ie

$$\bar{S} = \frac{f}{H - \bar{h}}$$

where $\bar{S}$ is the average scale, f is the focal length, H is the flying height above the datum plane and $\bar{h}$ is the mean height of the terrain.

The relief displacement can be employed to determine the height of an object because, with reference to Fig 2.24, it can be mathematically related to the height of point A thus:

$$\frac{\Delta r}{r} = \frac{h_A}{H} \text{ by similar triangles}$$

$$\therefore \Delta r = \frac{r \cdot h_A}{H}$$

$$\text{or } h_A = \frac{H \cdot \Delta r}{r}$$

where h_A is the height of the object A above the datum,
Δr is the amount of relief displacement,
r is the distance on the photograph from the principal point to the image of the top of the object (ie the radial distance), and H is the flying height from the datum.

50

This is the relief displacement formula which is particularly useful to the geographer who wants to determine the heights of a few prominent objects such as buildings, trees, etc, whose base is clearly visible on the photograph.[24] Obviously, objects near the edge of the photograph or at the furthest distance from the principal point give rise to larger relief displacements and hence their heights are much easier to determine. In practice, to obtain accurate measurement of relief displacement of an object, a scale magnifier is required.

The foregoing discussion on relief displacement assumes that the aerial photograph is perfectly vertical. When tilting is introduced, the relief displacement can be seen to be radial from the photo nadir point (n) (Fig 2.25).

THE COMBINED EFFECTS OF TILT AND RELIEF DISPLACEMENT

It has already been shown that tilting of the aerial photograph causes image displacement radial from the isocentre on a flat plane. With the presence of relief, the image is further displaced as a result of relief displacement caused by the central projection of the aerial photograph, which is radial from the nadir point. Thus, an aerial photograph is really the result of the interaction of these two processes. Moffitt has idealised their combined effects graphically as Fig 2.26 where the displacement of five points at various portions of the aerial photograph is examined.[25] In this diagram, the various positions of the point before and after tilting and/or relief displacement are indicated; thus number 1 indicates the correct position of the point at datum, number 2 indicates the position of the point after relief displacement, and number 3 indicates the position of the point after tilt displacement. It is noteworthy that for points A and B the tilt and relief displacements tend to act in opposite directions; for points C and D, the two displacements occur in the same direction; and for point E there is no tilt displacement (only relief displacement) because it lies on the axis of tilt. From this diagram, one can see the complexity of tilts and relief displacements in affecting the geometry of the aerial photograph.

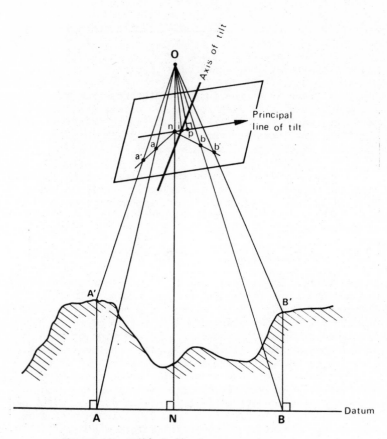

Fig 2.25 *Relief displacement in a tilted photograph*

ANHARMONIC PROPERTIES OF AN AIR PHOTOGRAPH

The aerial photograph, as has been indicated, is a graphic record of visible light rays through the medium of a light-sensitive emulsion coated on a stable base. As it is two-dimensional, only pencils of rays can be presented, which relate the photograph with the ground by means of projective geometry. As can be seen in Fig 2.27, the points A, B, C, D and E on the ground are imaged as a, b, c, d and e on the photograph. Thus, by

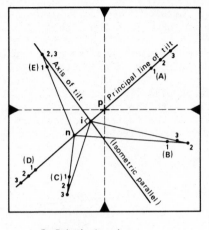

P = Principal point
i = Isocentre
n = Nadir point

Fig 2.26   *Combined effects of tilt and relief displacements*

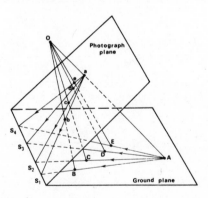

Fig 2.27   *Anharmonic properties of an aerial photograph*

virtue of the projection principles, *OaA* and *ObB* form two straight lines. Therefore, points *O*, *a*, *b*, *A* and *B* all lie in the same plane. Then the line *ab* produced will meet *AB* produced at S_1 on the line of intersection between the photograph plane and the ground plane. This applies to other points in relation to *OaA*.

If the photograph plane is turned about the line S_1S_4 until it lies flat with the ground plane, the projective relationship between the two can be seen more clearly (Fig 2.28). These projectively related figures possess

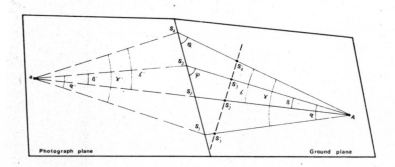

Fig 2.28 *Photograph plane having been rotated about S_1S_4*

a metric characteristic, called the *anharmonic* or *cross ratio*.[26] For Fig 2.28 the cross ratio is

$$\lambda_{(S_1S_2S_3S_4)} = \frac{S_1S_3}{S_2S_3} : \frac{S_1S_4}{S_2S_4}$$

By applying Sine Rule and substituting, it can be proved that

$$\lambda_{(S_1S_2S_3S_4)} = \frac{\sin\alpha}{\sin\beta} : \frac{\sin\gamma}{\sin\delta}$$

54

Thus, the cross ratio does not depend on the position of the cutting line and the distance along the cutting line such as S_1S_4, but depends only on the *angles* from which the rays come. Therefore, any cutting line can be drawn to intersect the pencil of rays in four points, such as $S'_1S'_4$ in Fig 2.28, which determine the value of λ. This ratio is very useful for photogrammetric mapping using a single aerial photograph.

GEOMETRY OF AN OVERLAPPING PAIR OF AERIAL PHOTOGRAPHS

In photogrammetry and photo-interpretation, overlapping pairs of vertical aerial photographs capable of generating three-dimensional views of the area concerned (ie stereomodels) are usually employed to allow the extraction of height data and a better appreciation of the terrain.

Parallax

The perception of distance by means of binocular vision depends on the magnitude of the parallactic angle (α) formed by the two eyes at the point of convergence. As shown in Fig 2.29, convergence of the axes of the eyes

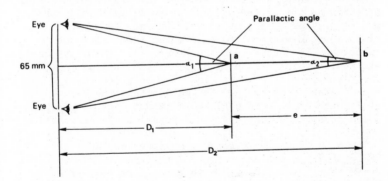

Fig 2.29 *Parallactic angles and distances*

at a gives rise to an angle α_1 at a distance D_1 and convergence at b gives rise to α_2 at D_2. The angular difference, ie $(\alpha_1 - \alpha_2)$, tells the mind that the distance between a and b is e. By analogy, for an overlapping pair of aerial photographs, the two exposure positions of the aerial camera represent the two big eyes of a giant looking down on to the ground, and by means of the differences in parallactic angles, the heights of various objects from the ground can be determined.

Thus, by having a pair of overlapping photographs taken at two different view points, *parallax* is produced. Parallax is therefore the apparent displacement of the position of a body, with respect to a reference point or system, caused by a shift in the point of observation.

There are two types of parallax, the X-parallax and the Y-parallax, according to the direction in which these parallaxes occur during the stereoscopic veiwing of the pair of photographs.

Y-Parallax, or Want of Correspondence, or vertical parallax, occurs as a result of (1) unequal flying height, (2) tilt in the Y-direction and (3) misalignment of the flight line.

For (1), when the flying heights for the pair of aerial photographs are different (but without any tilts), the sectional view of the photo-pair becomes as shown in Fig 2.30a, an exaggerated case which shows that one photograph is higher than the other, thus resulting in the scale of photo 1 being smaller than that of photo 2. The plan view is shown in Fig 2.30b in which the Y-coordinates of b_1 (ie y_b) is smaller than the Y-coordinate of b_1' (ie y'_b) by an amount Δy_b (ie $y'_b - y_b$). The discrepancy of Δy_b is the Y-parallax of the point.

The same kind of discrepancy measured in the Y-axis of the photograph can occur for (2) tilting of one photograph relative to the other in the Y-direction (Figs 2.31a and 2.31b) and for (3) misalignment of the flight line between two photographs (Fig 2.32).

It is obvious that the three effects of (1), (2) and (3) can be combined to produce even more complex Y-parallax, and can cause great difficulty in stereoscopic

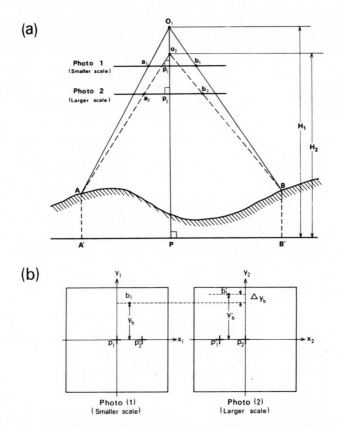

(a)

Photo 1
(Smaller scale)

Photo 2
(Larger scale)

(b)

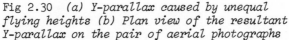

Photo (1)
(Smaller scale)

Photo (2)
(Larger scale)

Fig 2.30 *(a) Y-parallax caused by unequal
flying heights (b) Plan view of the resultant
Y-parallax on the pair of aerial photographs*

viewing of the pair of photographs.

X-Parallax is the apparent image displacement created
along the X-direction on the stereopair of aerial photographs.

To derive a geometrical definition, one may consider
the parallax of a point *A* in Fig 2.33, which can be
defined as the angle subtended at *A* by the projection
centre but also as linear distances measured on the

57

(a)

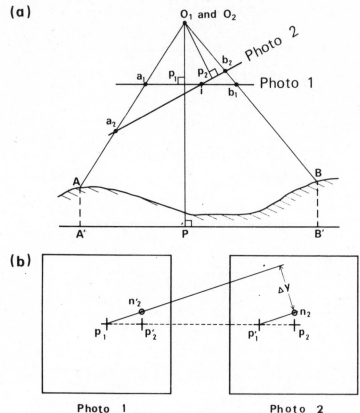

(b)

Photo 1   Photo 2

Fig 2.31 *(a) Y-parallax caused by the tilting of photo (2) relative to photo (1) in the Y-direction (b) Plan view of the resultant Y-parallax: since only photo (2) is tilted relative to photo (1) in the Y-direction, a nadir point (n_2) can be located on photo (2) as shown.*

photograph.

Thus, with reference to Fig 2.33, for the negative plane, the parallactic angle ($\bar{\alpha}$) becomes:

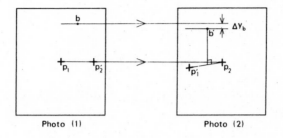

Fig 2.32 *Y-parallax caused by misalignment of flight lines between photos (1) and (2)*

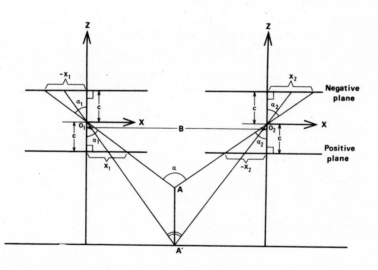

Fig 2.33 *Definition of X-parallax*

59

$$\bar{\alpha} = \alpha_1 + \alpha_2$$

$$\doteq \frac{-x_1}{c} + \frac{x_2}{c}$$

$$\doteq \frac{-x_1 + x_2}{c}$$

For the positive plane, the parallactic angle similarly ($\bar{\bar{\alpha}}$) becomes

$$\bar{\bar{\alpha}} \doteq \frac{x_1 - x_2}{c}$$

In general, since c in the formula above is the principal distance, which is a constant, the parallax may be more simply expressed as $\alpha = x_1 - x_2$ or $x_2 - x_1$ depending on whether the positive or negative plane is used. Thus defined, this is the algebraic difference in the X-coordinates of a point and is therefore known as the X-parallax, or linear parallax or horizontal parallax.

The X-parallax is usually employed for height measurement in photogrammetry according to the parallax formula derived below.

With reference to Fig 2.34, the geometric definition of the absolute stereoscopic parallax or X-parallax at the point $A(P_A)$ is

$$P_A = a'_2 p_1 + p_1 a_1 = a'_2 a_1 \text{ (since } O_1 a'_2 \text{ // } O_2 a_2)$$

Thus, by similar triangles $\Delta a'_2 O_1 a_1$ and $\Delta O_1 A O_2$,

$$\frac{a'_2 a_1}{c} = \frac{B}{H - h_A}$$

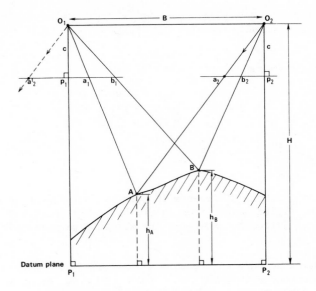

Fig 2.34 *Relationship between the height
of a point and its parallax for derivation
of the parallax equation*

∴ by substitution, $P_A = \dfrac{c \cdot B}{H - h_A}$

Similarly, for the point B, the absolute parallax, P_B, is

$$P_B = \frac{c \cdot B}{H - h_B}$$

The difference in parallax between the two points A and B
is therefore $P_{BA} = P_B - P_A$ since point B is higher than
point A (ie $h_B > h_A$).

By substitution,

$$\Delta P_{BA} = \frac{c \cdot B}{H - h_B} - \frac{c \cdot B}{H - h_A}$$

61

$$= \frac{c.B(h_B-h_A)}{(H-h_B)(H-h_A)} = \frac{P_B(h_B-h_A)}{(H-h_A)}$$

Let $\Delta h_{BA} = h_B - h_A$

$$\therefore \Delta P_{BA} = \frac{P_B \cdot \Delta h_{BA}}{(H-h_A)}$$

or $\Delta h_{BA} = \dfrac{P_{BA}(H-h_A)}{P_B} = \dfrac{P_{BA}(H-h_A)}{P_A+\Delta P_{BA}}$

where ΔP_{BA} is the difference in parallax between B and A; H is the flying height from datum; h_A is the height of the point A from datum; P_A is the absolute parallax for A.

This is the parallax formula and can be used, if the height of one point, say A, is known, to find the height of the other point B. The difference in parallax is measured by means of a simple instrument called the *parallax bar* or the *stereometer*, whilst the absolute parallax of the reference point A with known height must be measured from the pair of photographs according to the geometric definition of X-parallax given earlier.

It should be noted that this parallax formula has been derived assuming that the photographs used are perfectly vertical. If this is so, the heights computed by this formula should be correct. Unfortunately, this is not true in reality where tilts normally occur in photographs so that the resultant heights computed are crude heights only and need to be corrected for the *deformations* of the stereomodel when viewed through the mirror stereoscope. These model deformations are caused by the tilts (ϕ, ω, κ) and shifts (bx, by and bz) of the aircraft at the time of photography and can be separated into the following five types:

1 *Errors in bx or a shift along the X-direction.* This is a movement of the projection centre along the X-axis, thus giving rise to X-parallaxes and hence a change in

the height of the model plane (Fig 2.35a)

(a) Errors in bx

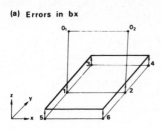

(d) Errors in ω

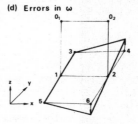

(b) Errors in bz

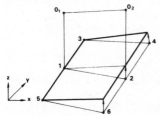

(e) Errors in φ

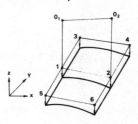

(c) Errors in Κ

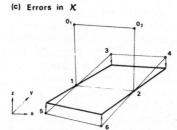

Fig 2.35 *Types of model deformation*

2 *Errors in bz or a shift in the Z-direction.* This is
the movement of the projection centre in the Z-direction,
resulting in one photograph being taken at a higher
altitude than the other. The effect is a longitudinal
tilt of the model (Fig 2.35b). This gives rise to errors
linear in the X-direction.

3 *Errors in κ or a rotation about the Z-axis.* The
effect is a transverse tilt of the model (Fig 2.35c) and
gives rise to errors linear in the Y-direction.

63

4 *Errors in ω or a rotation about the X-axis*. The
effect is a twisting of the model whose base line remains
unchanged (Fig 2.35d), and results in errors which are
functions of xy.

5 *Errors in φ or a rotation about the Y-axis*. The
effect is a concavity in the model plane (Fig 2.35e) and
gives rise to errors which are functions of x^2.

   The total error of these five kinds of deformation can
be expressed mathematically as

$$dh = a_0 + a_1 x + a_2 y + a_3 xy + a_4 x^2$$

where $dh = h'-h$ (in which h' is the crude height and h
is the correct height);
   $a_0,\ \ldots\ldots\ a_4$ are error coefficients which express the
magnitude of the rotation,
   and x, y are coordinates of these points on the model.

   The error coefficients can be obtained by solving five
simultaneous equations in the form shown above. This
requires five control points with known X-, Y- and Z-
coordinates to be distributed in the model as shown in
Fig 2.36(a). The method commonly used in the solution
of the five simultaneous equations is the Doolittle
Method which is rather tedious and requires great care
during solution.[27] This manual method of solution is
eased with the aid of tabulated forms and is fully
explained in Thompson[28] and Kilford.[29] A graphical
solution is also given if a large number of points are
required. Today, with the electronic computer becoming
more generally available, the procedure of solution can
be easily programmed as demonstrated by Methley.[30] One
great advantage in the use of the computer is the
flexibility made available by the program to undertake
the solution of even more complicated equations.
Accuracy of the heighting result involving the use of a
simple parallax bar can be greatly improved by employing
more than five control points, eg 7, 9, or 13 (Fig 2.36),
to take into account more complicated deformations of the
stereomodel. There appears to be an optimum number of
control points which can be used. Any number greater
than the optimum will not bring about any significant
improvements in accuracy.[31] It is also possible by means

64

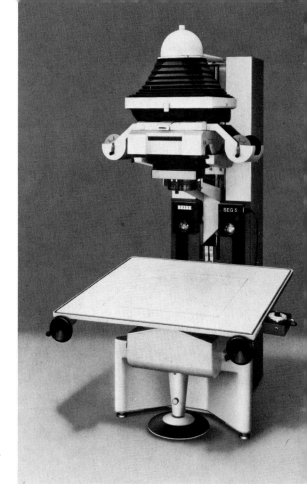

Plate 5 SEG-V Rectifier
(Carl Zeiss, Oberkochen)

Plate 6 Radial Line Plotter
(Rank Precision)

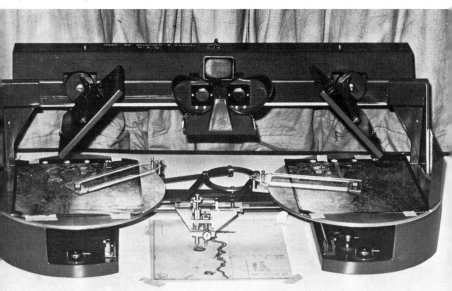

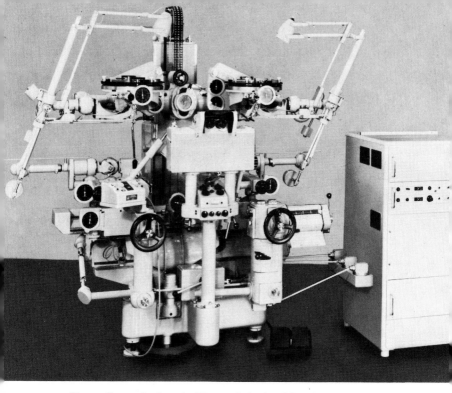

Plate 7 Stereoplanigraph C8 coupled with SG-1 Storage Unit *(Carl Zeiss, Oberkochen)*

Plate 8 Stereosimplex model IIc *(Officine Galileo)*

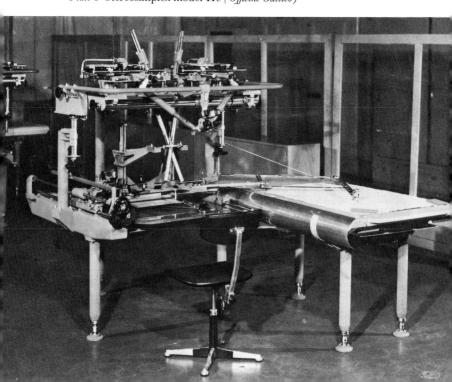

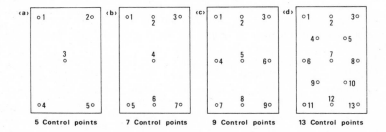

Fig 2.36 *Patterns of control point distribution*

of programming to take out excessive relief displacements in the stereomodel - a typical analytical approach.

PROVISION OF CONTROLS

The practical application of the principles of photo-grammetry in topographic mapping results in the development of instruments at varying degrees of sophistication for planimetric plotting and/or contouring. But before the instruments can be used for these purposes, it is necessary to establish the image-ground relationships and to join successive strips of photographs together. All these require control points, the provision of which constitutes a major phase in the whole process of photo-grammetric mapping.

There are different types of control point according to the purposes they are intended for, as follows.

1 *Ground Control Points*. Control points are required to be set up on the ground to relate the stereomodel and the terrain geometrically. These take the forms of (a) *the planimetric control point* which is the horizontal position of a point with respect to a horizontal datum (ie as *X*- and *Y*-values in a rectangular coordinate system) and (b) *the height control point* which is the

height of a point with respect to a vertical datum (ie Z-value in the same system). At least two planimetric control points are required to determine the azimuth and scale of the stereomodel, and a minimum of three height control point is needed to allow the stereomodel to be orientated with respect to the height datum for heighting. The whole procedure is known as absolute orientation.

The accuracy of these ground control points determines the resultant accuracy of the map produced from the stereomodel by the machine. Usually, these ground control points have to be actually surveyed in the field, but can be known triangulation stations if a surveying framework already exists. If no such suitable triangulation stations are found within the overlap, the planimetric control points are usually fixed by running a close traverse with a theodolite from a nearby triangulation station or other points of known coordinates, or by resection or intersection or trilateration, whilst the height control points are determined by carrying a line of levels from a point of known height (such as a bench mark) to the desired point with a levelling instrument. It is important that the accuracy of the ground control is compatible with the type of map to be produced. Thus, large-scale maps with small contour intervals should have more accurate planimetric and height controls than those for small-scale maps with large contour intervale. The accuracy of the ground controls point is therefore determined by the order of ground surveying that has been employed. For general purposes, it is adequate to fix the height control points to an accuracy of ±0.1X contour interval and planimetric control points are fixed such that they can be plotted to an accuracy of ±0.15mm at mapping scale.

There are different requirements for planimetric and height control points. For planimetry, the control points should be well defined with sharp contrast against their surroundings such as the junctions of hedges, fences, walls and corners of features such as footpaths; but for height, they should be located in open ground and flat areas of good photographic texture, such as the centres of level road or footpath junctions, and flat roofs of structures. Thus, a single control point will not usually be able to serve both purposes; hence the

66

two sets of control are selected independently.[32]

2 *Minor Control Points* (or Pass Points or Wing Points).
These are supplementary controls established on the photographs to serve two purposes: (a) to join the individual
models together to form a strip; and (b) for the absolute
orientation of each model. For the first purpose, they
must be located in the common overlap between two models,
ie on three successive photographs (Fig 2.37), and for
the second purpose, they must be chosen close to the
edge of the models so that good control in the common Ω
tilt is obtained. A third point in the centre of the
overlap is also taken.

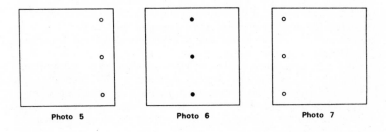

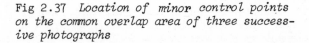

Fig 2.37 *Location of minor control points
on the common overlap area of three successive photographs*

3 *Tie Points.* These are supplementary control points
selected to connect strips of photographs together to
form a block. They are therefore located at the centre
of the lateral overlap between strips. Usually, one tie
point per model is taken, but for lower accuracy it will
be sufficient to have a tie point every alternate model.
If very high accuracy is required, two tie points per
model should be taken.

In some cases, a minor control point suitably located

at the edge of the photograph can be used also as a tie point. The optimum case arises when a minor control point can be found in six photographs, ie on three successive photographs in one strip and on three successive photographs in the adjacent strip.

As these three types of control points have to be accurately identified on the photographs, they are best signalised on the ground before flight (called Pre-marking).But this is not always possible; and normally only ground control points are *signalised points* because fewer of these are required and the highest accuracy needs to be maintained for them. On the other hand, it is also quite common to make use of *natural points* after flight for ground control points, minor control points and tie points. Natural details such as stones, bushes, road intersections, fence corners, etc which are clearly identifiable on the photographs, are chosen for this purpose. It is important to keep a good illustrated record of these natural points to help their recovery on the ground or pointings on the model. Since the minor control points and tie points need not be established in the field, they can be selected without regard for ease of access, etc, as in the case of ground control points, but only with regard for their usefulness. All these points, once selected, should be pricked carefully with a pin on the photographs and accurately transferred from one photograph to the other through stereoscopic viewing.

Aerial Triangulation

Apart from the ground controls,all other types of supplementary planimetric and height controls can be established photogrammetrically by a method called aerial triangulation, using the geometric relationships between successive aerial photographs, by means of which the points so established are tied to the ground control to give them geodetic significance. Therefore, with the use of aerial triangulation, the number of control points determined by ground survey methods is drastically reduced.

Essentially, there are two types of aerial triangulation:

1 *radial aerial triangulation* in which measurements are made on the plane of the photograph itself without considering the third dimension, thus giving rise only to planimetric controls (ie minor control points) as pairs of X- and Y-coordinates; and

2 *spatial aerial triangulation* in which the three-dimensional stereomodel is reconstructed, usually in an analogue instrument or by computation using an analytical approach, thus capable of giving to both height and planimetric controls (ie X-, Y- and Z-coordinates of a point).

1 *Radial Triangulation*. This method is based on the so-called *radial line assumption* which states that angles measured from the principal point in a near-vertical aerial photograph are equal to the angles at the terrain corresponding to the principal point of the photograph. This in fact is not true, because, as we have seen in an earlier section, displacements on an aerial photograph taken with a tilted camera axis are radial from the *isocentre* whilst those caused by the relief are radial from the *nadir point*. Such an assumption has been found to be responsible for some systematic errors in photo directions which are functions of tilts and the degree of variations of relief in the terrain.[33]

A *graphical* approach to this method of radial triangulation is quite commonly employed. A *mechanical* approach called *slotted template triangulation* is also possible (Plate 2). A full account of the procedures can be found in any photogrammetry textbooks, eg Kilford[34] and Moffitt.[35] Accuracy depends on the relief of the terrain and the amount of tilting present in the aerial photographs.

2 *Spatial Aerial Triangulation*. This is more complex than radial triangulation, since the third dimension is also involved. The analogue procedure requires the use of a slightly more sophisticated stereo-photogrammetric instrument equipped with a series of projectors by which stereomodels can be set up one after another so that individual models in one strip are connected together (Fig 2.38). The setting up of the first model requires proper relative and absolute orientation. Hence, the

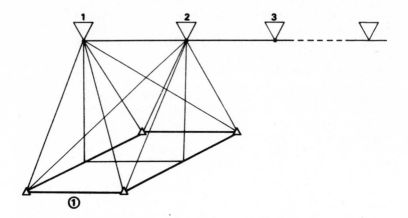

Fig 2.38 *Spatial aerial triangulation*

main principle of spatial aerial triangulation is the
transfer of the elements of orientation of the first
model from model to model. The actual procedures involved vary from one type of machine to another. Thus, the
Aeropolygon Method is generally employed for Multiplex-
type and universal machines, whilst the *Method of Independent Models* is favoured for a precision or topographic
plotter.

Basically, the result from spatial aerial triangulation
is a list of coordinates of points of interest in the
coordinate system of the first model in a strip. Discrepancies (called 'cracks') may occur between successive
models, which produce greater and greater errors as one
model after another is connected. A statistical adjustment has to be carried out to eliminate these discrepancies. The accuracy of aerial triangulation by the
method of independent models carried out on topographic
plotter is about 0.2m at model scale after adjustment.

An analytical method of spatial aerial triangulation
involving the use of the digitiser and the computer is
also possible, which considerably speeds up the whole
operation.[36]

PHOTOGRAMMETRIC PLOTTING INSTRUMENTS

Different types of topographic mapping instruments have
been developed based on the principle of stereo-photo-
grammetry where vertical aerial photography is utilised.
These range from the simplest to the highly sophisticated.
For convenience, one may divide them into two classes
based on their functions: (1) instruments designed for
plotting planimetric detail, and (2) stereoplotting
machines.

1 *Planimetric Plotting Instruments*

The instruments under this class have been designed only
for transferring planimetric details directly from a
single aerial photograph, or a stereopair, to paper (or
map). Among these, the following possess special potential for geographical applications:

a *Sketch Master*. An example is the LUZ Aero Sketchmaster manufactured by Zeiss (Oberkochen)(Plate 3). This
uses only a single aerial photograph and does not give a
stereomodel. Basically, the instrument consists of three
components: the photograph carrier, the map sheet carrier
and the viewing device. Through the special viewing
device (which is made up of two prisms) both the photograph and the map can be seen as superimposed. The
effect of tilts present in the photograph can be taken
out by moving the photograph corner about the X-, Y- and
Z-axes, and the scale difference between the map and the
photograph itself can be adjusted by varying the photo
and map distance. Since the viewing device enables only
a limited area of the whole photograph to be seen, the
photograph has to be divided into smaller areas of
uniform relief during plotting. Thus, the Sketch Master
is a convenient map revising device. Because of its
power to remove tilts from the photograph by tilting and
turning the photograph carrier, the Sketch Master may
also be regarded as a subjective projection rectifier
using the anharmonic properties of the aerial photograph.
A more recent development of the Sketch Master idea is
the production in the USA of a Zoom Transfer Scope by
Bausch and Lomb (Plate 4). The main feature of the
instrument is a zoom stereoscope which allows the whole
photograph to be viewed comfortably at one glance whilst

a more detailed view can also be obtained for a small area by zooming the stereoscope binocular. The optical system allows shape transformation to be made so that individual features as seen from the photograph can be fitted properly to the corresponding features on the map when superimposed, and any scale change can also be easily carried out. By obtaining the 'best fit', much of the relief displacements and photo tilts can be eliminated.[37] Such an instrument is in fact carrying out rectification of the photograph optically because rectification, by definition, is a process aimed at 'transforming one photographic image into another which has the same planimetric properties as a map, including a known scale'. This instrument is therefore particularly useful in viewing satellite photographs where serious shape distortions inevitably occur owing to tilting and other sources of error.

b *Optical Rectifier*. More usually, rectification of the aerial photographs is carried out on the *optical rectifier* (or *objective projection rectifier*). The optical rectifier consists of a projector and a flat board which is free to move about the X- and Y-axes. The condenser in the projector directs the rays of light through a negative or a diapositive and then through a lens to focus on to the flat board. The optical recti- fier can maintain a correct relationship among three planes: (a) the negative plane, (b) the lens plane, and (c) the easel plane or map plane. It is important for the rectifier to maintain the sharpness of the image, for which the Scheimpflug condition needs to be satis- fied. The Scheimpflug condition states that the negative plane, the lens plane and the easel plane must intersect along one line (Fig 2.39), which is equivalent to satisfying Newton's Lens Equation, ie with the notation in Fig 2.39, $1/a + 1/b = 1/f$ where f is the focal length of the lens.

The projected image is rectified on the easel plane (ie the board) with reference to at least four control points by tilting the board to remove the photo tilts, after which a print of the rectified image at any desired scale can be made - ie a rectified photograph results. Optical rectification forms the basis for automatic or semi-automatic production of orthophoto-

72

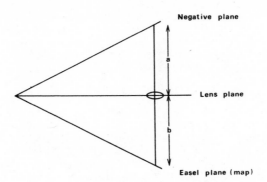

Fig 2.39 *Scheimpflug condition*

graphs. A notable example of the optical rectifier is
the Standard SEG V Aerotopo-Carl Zeiss (Oberkochen)
Rectifier (Plate 5).

c *Radial Line Plotter*. Planimetric plotting, however,
is more easily carried out over a large area using a
stereopair of aerial photographs which give rise to a
stereomodel. The radial line assumption as used in
radial triangulation can be employed for this purpose so
that an analogy of the plane table survey method can be
carried out on the aerial photographs. An instrument
especially suitable for this type of application is the
Radial Line Plotter (Plate 6). This consists of (a) a
mirror stereoscope; (b) two photograph carriers; (c) two
cursors made of transparent material (with a narrow slot
cut and a straight wire fitted taut in the middle) which
pass above and below the photograph carriers and rotate
about the centres of the photograph carriers (ie marking
the two principal points); (d) a parallel bar which moves
parallel to the photograph base and the ends of which
connect to the lower part of each radial arm by means of
a pin fitted into a slot in the arm; and (e) a parallel
guidance system connecting the parallel bar to a pencil
holder (Fig 2.40). In operation, the two cursors represent two radial rays drawn from the principal points of
the respective photographs in the direction of a certain

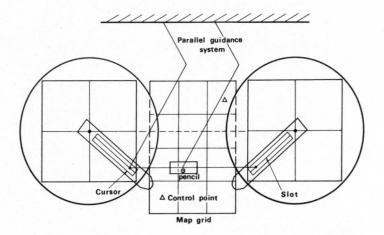

Fig 2.40 *Principle of the radial line plotter*

feature. When viewed through the mirror stereoscope, the position of this feature can be fixed by the two radiating lines. Correct orientation and scaling of the stereomodel need to be made first with respect to the ground control points already plotted on the map grid which is placed directly under the instrument on the table. A slight scale adjustment is possible (from ±0.5 to 2 times). After this, individual features can be intersected by the two cursors as seen through the mirror stereoscope, and continuous lines of features (eg rivers and roads) can be broken down as discrete points and intersected to be joined up afterwards. For features near the photograph base, eccentrically placed radial centres can be established directly above and below the principal points, over which the two cursors can be pivoted for intersection. The features intersected in this region are usually less accurate. But, on the whole, this instrument gives reasonably accurate plans, and is particularly useful to geographers in transfer-

ring up-to-date details from the photographs on to the
base maps.

2 *Stereoplotting Machines*

Planimetric plotting from single aerial photographs or
stereopairs is not always sufficient; and a knowledge
of heights (ie the third dimension) is usually required,
especially when topographic maps are to be drawn.

The stereoplotting machine performs exactly the same
function as the mirror stereoscope and the parallax bar -
ie to create a stereomodel and to accurately measure the
X-parallax by means of the floating mark method. In
addition to this, the movements of the floating mark can
be transferred to the plotting pencil of a pantograph so
that contours and planimetric details can be directly
plotted from the stereomodel. The solution of the
parallax equation for heights is carried out in the
machine by an analogue process, and the stereoplotting
machine itself can be regarded as an analogue computer
since it mechanically simulates what happened in the air
at a reduced scale so that the correct stereomodel can be
reconstructed. The scale of this stereomodel is determined by the separation of the two aerial photographs (ie
the length of the air base).

The creation of the stereomodel can be achieved by
either *optical* or *mechanical* means. By optical means, a
projecting lamp is used for each photograph so that both
the left and right diapositives are projected directly
on to a common plane for viewing. The result is that
those parts of the stereomodel at the edges of the
diapositives are seen more obliquely than those near the
centre. A common viewing system for the optical projection machines is the anaglyphic system which involves
the projection of the diapositives through red and green
filters under darkroom conditions and the viewing of the
stereomodel through a pair of spectacles with green-red
filters. The scale of the stereomodel resulting depends
not only on the original scale of the diapositives used
but also on the distance of the projectors from the
projection surface. Examples of this type of stereoplotting machine include the Multiplex, the Kelsh Plotter
and the more recent Zeiss (Oberkochen) DP-1. An example

of the optical projection machine not using anaglyphic viewing is the classic Stereoplanigraph manufactured by the firm of Zeiss in both East and West Germany (Plate 7).

As for the mechanical means of projection, light rays are replaced by metal rods in the stereoplotting machine. The exact relation between a point on the photograph and the corresponding point on the ground (Fig 2.41) is

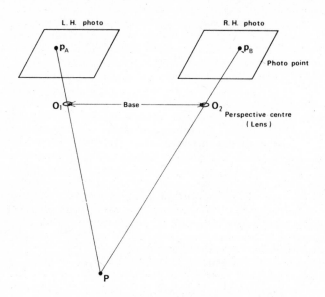

Fig 2.41 *Geometric relationship between a point on the photograph and the corresponding point on the ground*

duplicated mechanically using gimbals and space rods as shown in Fig 2.42. The perspective centres (O_1 and O_2) are fixed points represented by gimbals through which the space rods will always pass. Thus, these points are regarded as universal joints. The space rod is connected to a small microscope which moves with the rod in such a way that it allows any part of the photograph to be seen orthogonally (called frontal observation system). Examples of machines constructed on this principle

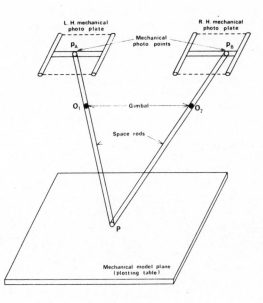

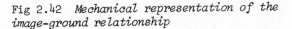

Fig 2.42 *Mechanical representation of the image-ground relationship*

include Zeiss (Jena) Stereometrograph; Zeiss (Oberkochen) Planimat; Galileo-Santoni Stereosimplex I, II, III (Plate 8); Wild A8, B8S (Plate 9); Kern PG-2 and PG-3 (Plate 10).

Setting up Procedures

To recover the stereomodel correctly, the following procedures need to be carried out in any plotting machines:

1 *Inner Orientation*. This involves correct centring and setting the correct principal distance of the aerial camera on the machine. These together will ensure the same bundle of rays at the time of exposure for the photographs to be recovered exactly.

2 *Relative Orientation*. This is to provide the correct orientation between two photographs. When the photographs were exposed, tilts also occured which must

77

be correctly set in the plotting machine. It has been
seen that any tilts present can be duplicated by three
rotations about the three axes (ie Omega (ω), Phi (ϕ),
and Kappa (κ)) and three shifts (ie bx, by, and bz
caused by variations in flying height along the three
axes). Thus each projector on the machine has six types
of movement (known as degrees of freedom) and two projectors together give twelve. The stereoplotting machine
is equipped with these *orientation elements*. Usually not
all twelve but only a minimum of five are required, since
the effect of the remaining elements can be duplicated
by the others.

Thus, relative orientation in actual practice requires
the elimination of the Y-parallax at five standard
positions on the overlap (equivalent to the solution of
five simultaneous Y-parallax equations by computation)
whilst the sixth position is used as a check (Fig 2.43).

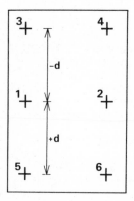

Fig 2.43 *The six standard positions for
relative orientation by an analogue plotter*

The effects of movements of the six orientation elements
are shown graphically in Fig 2.44 where nine instead of
six positions are used. The standard practice of
relative orientation varies from one machine to another,
depending on the types of orientation element available.

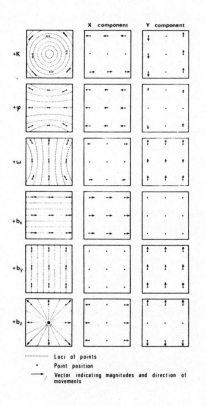

| X component | Y component |

+K

+φ

+ω

+b_x

+b_y

+b_z

------ Loci of points
• Point position
⟶ Vector indicating magnitudes and direction of
movements

Fig 2.44 *Effects of movements of orien-*
tation elements on a single projected
image (after van der Weele, 1966)

For details of the procedures, one should consult the
instrument manuals or any standard photogrammetry textbook such as Moffitt[38] and Wolf.[39]

3 *Absolute Orientation.* After relative orientation,
a parallax-free stereomodel is established, but this is
still not yet the correct model since it may occur at any
altitudes. for the bundle of rays may intersect at a
wrong altitude for example, on an inclined plane.

In order to fix it at the proper height, the whole model has to be rotated about two axes - first about the X-axis giving the common omega (Ω) rotation and then about the Y-axis - giving the common phi (ϕ) rotation as shown in Fig 2.45.

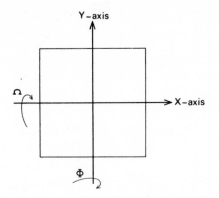

Fig 2.45 *Absolute orientation: rotation of the whole stereomodel about the X- and Y-axes*

In carrying out this procedure, the model must first be scaled, which simply involves changing the separation of the two projectors with reference to two control points marked on the map grid placed on the plotting table. A third control point is usually used to provide a check for scaling and for the subsequent absolute orientation.

After the completion of these three main procedures, direct plotting of contours and planimetric details from the aerial photographs can begin. The stereoplotting machine is characterised by a measuring system which measures in varying degrees of accuracy X-, Y- and Z-coordinates of each point on the model. This measuring system is directly connected to the floating mark, the movement of which in the X-, Y- and Z-directions can be measured by means of shaft counters. In some lower order plotting machines, such as the Kern PG2 and the Wild B8S, only the Z-value (ie heights) can be measured, thus

providing what is called a *free-hand motion*, and the
pointing of features as a result is less precise. The
contour is plotted continuously by keeping the floating
mark in contact with the ground at a pre-set height.
Similarly, the planimetric details are also plotted,
usually on a separate sheet, by means of the floating
mark following the outlines of individual features.

Classification of Stereoplotting Machines

Stereoplotting machines fall into very distinct groups
according to the number of functions that they can
perform and the degree of accuracy that is achievable.
One classification suggested by Professor W. Schermerhorn
of the International Institute for Aerial Survey and
Earth Sciences (ITC) in Enschede has been quite commonly
used. His scheme classifies the stereoplotting machines
into four types:

1 *Universal Machines* which are capable of performing
a multitude of functions, including large-scale plotting,
aerial triangulation, coordinate registration and
terrestrial photogrammetry. Examples include the Wild
A7, Galileo Stereocartograph V, Zeiss Stereoplanigraph C8,
Nistri Photostereograph Beta 2.

2 *High Precision Plotters* which can be employed for
large-scale plotting, aerial triangulation by the method
of independent models, and coordinate registration.
Examples are the Wild A8, A10; Zeiss (Jena) Stereometrograph; Zeiss (Oberkochen) Planimat; SOM Pressa 224;
Thompson-Watts Plotter; and Kern PG-3.

3 *Topographic Plotters* which can be employed for
medium- to small-scale plotting, aerial triangulation
by independent models and coordinate registration.
Examples are the Kern PG-2, Wild B8S, Galileo-Santoni
Stereosimplex IIc, Multiplex, Bausch and Lomb Balplex,
and Nistri Stereoplotter RA/II.

4 *Approximate Machines* which are machines only capable
of approximate solution of the parallax equation. They
can allow a little enlargement but the resultant accuracy
of the plot is relatively low. Examples are the Zeiss
(Oberkochen) Stereotope, Galileo-Santoni Cartographic

Stereomicrometer, SOM Stereoflex, and Cartographic Engineering CP-1.

For applications in geography, the approximate machines have been used more often because of their relative ease of operation and, most important, because they are relatively cheap. But their use is rather limited because only a small amount of enlargement in the original photograph scale is permissible, and because the heights obtained are rather poor in accuracy.

In the 1960s, a new class of machines at a lower price range which makes use of a rigorous solution of the parallax equation emerged.[40] These are the Topographic Plotters. Their major limitation is that at most only a two-time enlargement of the original photograph scale as the plotting scale is possible whereas in the Universal and Precision Plotters a large enlargement is possible. Specific examples of topographic plotters are the Galileo-Santoni Stereosimplex IIc (Plate 8), the Wild B8 Aviograph, now called the B8S (Plate 9) and the Kern PG-2 (Plate 10), of which the latter two have been most commonly employed by geographers today.

Finally, one's attention should also be drawn to a new machine, the Cartographic Plotter CP-1 designed by Professor E.H. Thompson of University College, London.[41] This machine has already been put into the approximate solution machine class because it does not carry out a rigorous solution as in the case of the topographic plotter. It appears, however, that this machine lies somewhere between the topographic plotter class and the approximate machine class because the instrument design was based on first-order formulae and hence it can achieve a higher accuracy. Thompson gave the standard error of heighting of about 0.2 per cent of the flying height for wide- and super-wide-angle photography (Plate 11). The machine consists of two distinct parts: the first part to provide a means of removing the effects of relief variations and to convert parallaxes to heights; the second part to ensure correction for tilt of the X- and Y- coordinates on each photograph independently - a mechanism which can carry out *rectification* of each photograph independently. The plotting has to be done with the use of a pantograph which permits enlargement from 0.5 to 3 times. The operation of the CP-1, however, is more complex and less conventional, which may be regarded as a great disadvantage.

Analytical Plotters

To complete this introductory discussion of stereoplotting machines, one needs to note the modern interest in analytical photogrammetry made possible by the more general availability of electronic computers. In analytical photogrammetry, a mathematical model is constructed to link up (a) points in the object space; (b) the perspective centre in the lens; and (c) the images on the photographs. The computer is used to carry out the solution of problems of camera calibration, space resection and orientation, space intersection, and triangulation of groups, strips, or blocks of aerial or terrestrial photographs.[42] This is really the establishment of the relationship between the photograph (image space) and the terrain (object space) coordinate systems; and for such a purpose, a special instrument called the *Stereocomparator* has to be used. This allows coordinate values (X- and Y-values) for points to be measured precisely (up to six digits). Examples are the PSK-Aerotop Zeiss (Oberkochen) Stereocomparator, the former Higer-Watts Recording Stereocomparator, Zeiss (Jena) Stecometer, etc (Plate 12). For subsequent analysis, the output of coordinate values can be directly recorded in tape for direct input in the computer for processing.

CONCLUSIONS

In this chapter, a brief survey of the theories and practices of photogrammetry has been given to provide the necessary background for understanding the ensuing discussion of the application of aerial photography to geographical studies. After some technical discussion of aerial photographic procedures and film types, the geometric relations between the photograph and the terrain have been explained and the fact that these can be exploited for topographic mapping by using either single aerial photographs or stereoscopic pairs of aerial photographs has been noted. Some commonly used instrumental methods have been explained. The principles and methods of aerial triangulation have also been discussed with reference to the need to establish the correct image-ground relationship for plotting. The important role played by the analogue-type stereoplotting machines in producing topographic maps at different scales quickly and accurately has been examined, and led to a detailed discussion of the class of topographic plotters as a modern geographic research tool. Attention has also been drawn to the analytical approach which will have a great impact on the future development of photogrammetry.

III AERIAL PHOTOGRAPHS AS METRIC MODELS

It has already been well established that the major application of aerial photography was and still is in the field of topographic mapping. Indeed, by employing a suitable instrument and a sufficient number of control points, a correct stereomodel can be reconstructed which can then be scanned and measured by means of the floating dot for the extraction of metric data and for the continuous plotting of contours at pre-determined intervals. One should note, however, that the accuracy of the data extracted and of the map produced depends on the scale of the photographs and the reliability of the control points used. In all these respects, the geometric properties of the stereomodel have been exploited. Thus, the resultant stereomodel may be regarded as a metric model of topographic and non-topographic information.

An examination of the geographical applications of aerial photography has shown that this vital aspect of the stereomodel has been largely neglected, especially by human geographers. Indeed, even by physical geographers who have paid more attention to the quantitative aspects of aerial photography, the potentialities of the metric characteristics of the stereomodel have not been fully exploited. This largely stems from the fact that geographers have been rather reluctant to make use of the more advanced photogrammetric plotters, partly through lack of training in the operation of these machines and partly because of the rather high capital investment involved. However, a review of the efforts already made by geographers in extracting metric data from aerial photographs will show up certain distinctive approaches.

EXTRACTION OF NON-TOPOGRAPHIC METRIC DATA
RELATING TO THE TERRESTRIAL ENVIRONMENT

The terrestrial environment has its natural and artificial components, the relations of which to man are the
objects of investigation by geographers. In dealing with
the natural environment, one is concerned with the
earth's crust, the natural vegetation cover and the
surrounding atmosphere. As for the artificial environment, the main concern is with the built forms, structures, and other landmarks left behind by man. Aerial
photography has been applied with varying degrees of
success to each of these elements of the terrestrial
environment for the extraction of quantitative information. The following exemplifies some of the more typical
approaches.

Geology and Geomorphology

The use of aerial photographs in geology and geomorphology forms a distinctive field which may be separately
identified as *photogeology and photogeomorphology*. [1]
In geology, one is interested in stratigraphy (ie the
arrangement of stratified rocks, and the fossils and rock
structures they contain), lithology (ie the rock types
and rock units) and geological structures. In geomorphology, the aim is to analyse land forms found on the
earth's surface, which involves breaking down the land
form into the basic unit of slope or an aggregate of
slopes. [2] These two branches of study are so closely
interrelated in their respective investigations that they
may be regarded as one, especially when the regional
geology of an area is to be dealt with.

For the quantitative approach, accurate metric data on
the various landform units are required, which can be
extracted from aerial photographs through measurement.
There are three distinct sets of parameters which are
directly obtainable: (1) lines and areas; (2) directions
and (3) heights. [3,4]

1 *Linear and Areal Data*. Linear data include stream
lengths; lengths of structural weaknesses such as faults,
joints and fractures; the length of a watershed or the
perimeter of a drainage basin, etc, which may be directly

measured from aerial photographs with the use of a scale
or a chartometer (ie a line follower) depending on
whether these linear features under measurement are regular or not. Minor linear features are best measured on
blown-up photographs. Obviously, the accuracy of such
measurements is limited by the degree of tilt and the
amount of relief displacement present in the photograph.
The areas will be over-estimated if they are found above
the datum and under-estimated if they occur below the
datum. Such errors can be neglected if these differences of altitude in the area relative to the flying
height are small. However, more accurate results can be
obtained if rectified photographs are used, or by using
the radial line plotter if ground controls are available
to permit setting up of the stereomodel at the correct
scale. The courses of the various linear features can
first be plotted from the photographs by intersection on
a plastic drafting sheet and then measured afterwards.

Areal data are usually obtained for basin areas and can
be directly made on the aerial photographs by means of
the polar planimeter, although again the measurement will
be subjected to scale error caused by relief changes and
tilts. A correction factor has to be included. It is
also possible to obtain the area by manipulating linear
measurements alone according to the appropriate formula
for a particular shape, for example, the use of Simpson's
Rule for irregular shapes. A more general approach is to
digitise the boundary of the enclosed area, ie to give
coordinate values for a series of points which best fit
the area.

Combinations of simple primary measurements, such as
the linear and areal measurements, can give rise to
useful ratios as secondary measures which are indicative
of the landform characteristics. A good example is
drainage density which is defined as the total length of
streams within an area divided by the area. This ratio
is considered to be closely related to the lithology or
the nature of the rock found in the area. Research
undertaken by Ray and Fischer in the US Geological Survey
found that circular sample areas gave more consistent
determinations of drainage density for any one rock type
within any one given area than did samples of small
individual drainage basins, and that the measurements

were inconsistent when photographs of different scales were used.[5] The latter discovery is obvious, as the ability to see small drainage rills decreases as the scale gets smaller. Despite this fact, a general agreement in the slope of the line for a rock type exists irrespective of the scale of the photographs used, as shown in Fig 3.1. Thus, it is possible to relate

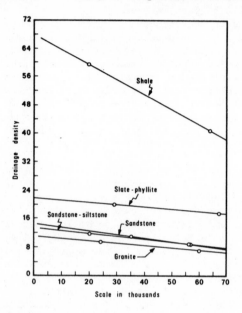

Fig 3.1 *The relationship of scale of aerial photographs to drainage density of several different rock types (after Ray and Fischer, 1960)*

drainage density with the scale of the photograph together so that a correcting factor can be computed. A general conclusion that emerged from such a study of drainage density was that the coarse-grained intrusive rocks show low drainage densities whilst the fine-grained clastic sedimentary rocks show higher drainage densities. Clearly, the permeability of the rock has greatly affected its drainage density.

2 *Directional Data.* The trends of the various linear features in a two-dimensional plane are also of interest to geologists, especially structural geologists. These are measurable in terms of the angular distances clockwise from a North-Point, known as bearings or azimuths. Also, the frequency of occurrence in different directions of these linear features is of interest. This is the method of *fracture analysis*, for which aerial photographs have been found to be particularly suited. The frequency of occurrence can of course be directly counted from the aerial photographs, but the azimuths of the linear features as measured directly from the photographs are not entirely correct because of relief displacement. Clearly, as shown in Fig 3.2, the azimuth of a line *de* will have

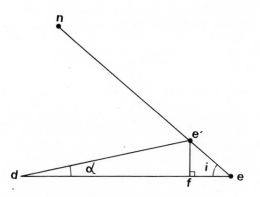

Fig 3.2 *Determination of the error in azimuth as measured from an aerial photograph*

an angular error α according to the amount of relief displacement *ee'* which is radial from the nadir point (*n*). The angular error (α) can be trigonometrically determined as follows:[6]

$$\tan \alpha = \frac{e'f}{df} = \frac{e'f}{de-ef} = \frac{ee'.\sin i}{de-ee'.\cos i}$$

since *ee'* = the amount of relief displacement which has

been proved in Chapter II to be equal to $\frac{r\Delta h}{H}$ where $\Delta h = H.\frac{\Delta P}{P+\Delta P}$

By substitution:

$$\tan \alpha = \frac{r.\Delta P.\sin i}{(P+\Delta P)(1-\frac{r.\Delta P.\cos i}{P+\Delta P})}$$

$$= \frac{r.\Delta P.\sin i}{(P+\Delta P) - (r.\Delta P.\cos i)}$$

The correction may be positive or negative depending on the position of the points d and e with respect to the nadir point on the aerial photograph.

The application of this formula is only suitable for obtaining the true azimuths of a few lines on the aerial photographs using the protractor. But for more practical purposes where a large number of azimuths has to be determined, as in the case of fracture analysis, a quicker method of obtaining the true azimuths has to be used. Blanchet described an example in which the Wild RC5 or RC8 camera with a focal length of 152mm and a distortion-free lens was used, giving rise to photographs of high resolution.[7] The method of slotted template was used to control the block of photographs covering the area for the construction of a controlled mosaic. Each photograph was then interpreted for fractures which were marked on a transparent sheet placed on top of the mosaic. In areas of excessive relief, all fractures which are not radial from the photo centre (ie the principal point) have to be azimuth-corrected by using a radial line plotter. After all these preliminary preparations, the azimuth (bearing) of each fracture was measured by means of the protractor, its length measured and the coordinates of its centre recorded (Fig 3.3). Analysis was then made to group the fractures together into sets, and fracture-azimuth frequency diagrams were plotted for each fracture set identified (Fig 3.4). Another fracture analysis carried out by Lattman and Nickelsen followed a more or less similar approach with the use of simple radial triangulation or a radial line plotter to determine the true positions of the ends of each fracture trace.[8] Small-scale aerial photographs

90

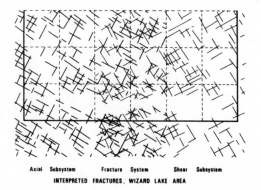

Axial Subsystem   Fracture System   Shear Subsystem
INTERPRETED FRACTURES, WIZARD LAKE AREA

Fig 3.3 *Interpreted fractures, from aerial photographs, of Wizard Lake Area, Alberta, Canada (from Blanchet, 1957)*

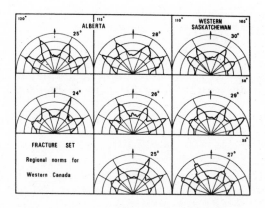

Fig 3.4 *Fracture-azimuth frequency diagrams for eight areas in Alberta and Western Saskatchewan, Canada. Note the general consistency in direction throughout the whole region (after Blanchet, 1957)*

(scale 1:20,000) were used, and they showed that in aerial photographs whose tilt is less than 3° and which cover areas of low to moderate relief, the errors in fracture-trace directions may be ignored if mapping on each photograph is not carried beyond an area whose sides are at half the distance from the photo centre to the four photo edges. More recent studies all confirmed and stressed the importance of such linear features as an aid to photo-geological research.[9] Stephens demonstrated the application of such a method of structural analysis in conjunction with the stereographic net to map structural trends over very large areas of regionally metamorphosed rocks and to delineate structural units.[10]

3 *Height Data*. Height data from aerial photographs are particularly important since the third dimension of the stereomodel is being measured. These include such measurements as height differences, slopes, dips, formational thicknesses and structure contouring, of which the slope measurement is particularly important.[11] All these types of measurement involve the use of the parallax bar or a wedge and the application of the parallax equation in the form given in Chapter II:

$$\Delta h = \frac{\Delta P(H-h)}{P+\Delta P}$$

For slope measurements on aerial photographs, different approaches are possible, as illustrated by Mekel, Savage and Zorn.[12] In general, the measurement can be based on pure computation using the definition for a slope as an expression of the steepness of the terrain which is, with reference to Fig 3.5,

$$\tan \theta = \frac{\text{vertical difference (V.D.)}}{\text{horizontal distance (H.D.)}}$$

or as its reciprocal trigonometric function:

$$\cot \theta = \frac{\text{horizontal distance (H.D.)}}{\text{vertical difference (V.D.)}}$$

The slope can be expressed as angles or in percentage by multiplying the ratio by 100 per cent. The parallax difference between the upper and lower points of the

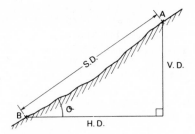

Fig 3.5 *Diagrammatic representation of a slope*

slope can be measured by means of the parallax bar and then converted into height difference by the parallax formula using the principal point of the photograph as the reference plane. The horizontal distance should be obtained through radial triangulation or by means of a radial line plotter, both at the photo-scale, if controls are available. The same approach can be used to determine the gradient of a river, for which a chartometer or line measurer has to be used to measure the total length of the river course.

A quicker but less accurate method of determining slope is to use the trigonometrical relationship $\sin \theta = \frac{V.D.}{S.D.}$ so that there is no need to determine the horizontal distance beforehand.

A more direct and accurate instrumental approach in slope measurement is to make use of a specially designed template in combination with the parallax bar, such as the ITC-Zorn method, which is capable of giving an accuracy of 2^{O}-4°.[13] The template is constructed based on the geometric principle of slope determination in which the horizontal distance between the two photo points on the slope is expressed as coordinates on a terrestrial system (U, V). The final formula becomes

$$\cot^2 \theta = (\frac{P_a}{\Delta P_{ba}} \cdot \frac{V_a - V_b}{C} + \frac{V_a}{C})^2 + \frac{U_p^2}{C^2}$$

93

where P_a is the absolute parallax for point A,
P_{ba} is the difference in parallax between A and B;
V_a is the V-coordinate of point A on the $U-V$
terrestrial coordinate system;
C is the principal distance
and U_p is the U-coordinate of the principal point on
the $U-V$ coordinate system.

The photo and terrestrial coordinate systems are related
as shown in Fig 3.6 where one should note that the V-axis

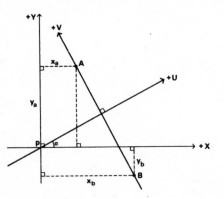

Fig 3.6 *Relationship between the
terrestrial and photograph coordinate
systems*

passes through the two points A and B and the U-axis
(which is parallel to the strike) passes through the
principal point p. The template (Fig 3.7) facilitates
the computation of the above equation by supplying values
for V_a, V_b and U_p on a special scale, on which the unit
is equal to C. These are given as the quantities
$t_a = V_a/C$, $t_b = V_b/C$ and $S = U_p/C$. The first two
quantities can be directly read from the template according to the appropriate principal distance of the aerial
camera used. The third quantity has to be obtained by
reading U_p, which is then divided by C. These values
obtained are substituted into the formula together with
the parallax measurement to obtain the correct angle of
slope. To facilitate this operation, a nomogram may also

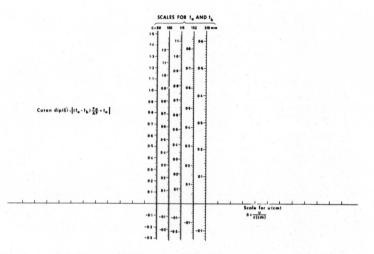

Fig 3.7 *Template produced by ITC for dip or slope measurement on vertical aerial photographs (after Mekel and Zorn, 1967)*

be used.

Other instrumental methods of slope measurement include the following:

a *The 'dipping platen'*, ie the tracing table used in conjunction with a double-projector Multiplex-type stereoplotter. The dip of the platen of the tracing table can be made to coincide with the ground surface or to intersect three or more points on a bedding plane when the stereomodel is viewed through the special spectacles. A clinometer is attached to the platen so that the angle of dip can be measured. But only moderate dips up to 40° can be measured by this method; for steeper dips the template methods should be used.[14]

b *Stereo Slope Comparator*. This is a simple instru-

95

ment developed by Hackman of the US Geological Survey for
use with paper prints.[15] It is essentially a parallax
bar except that each measuring dot is replaced by a
horizontal shaft at the end of which is mounted a small
target. The two targets can be fused stereoscopically to
give a single target (Plate 13). The horizontal shaft
can be positioned at any direction of strike and the
target at its end can be tilted to coincide with the
angle of dip. The protractor attached will measure the
dip of the target. It is noteworthy that only an exaggerated dip of the imaginary slope is measured since
the stereomodel has not yet been correctly reconstructed.
The reduction of the apparent dip to true dip can be
obtained by using the following relation:

tangent of true angle of dip

$$= \frac{\text{tangent of the exaggerated angle of dip}}{\text{exaggeration factor}}$$

The exaggeration factor is slightly different from one
person to another. A graph can be constructed to
facilitate the conversion from the exaggerated dip to the
true dip.

The determination of slope is obviously closely related
to the determination of dip and strike of sedimentary
rock strata, because the slope can be a dip slope. The
situation is best illustrated in Fig 3.8. The use of
parallax bar is essential. But it is necessary to know
the strike direction of the bed before the dip can be
measured. If this is not clear, two points of equal
parallax reading are selected on the dip slope so that
the strike can be indicated. All the methods of slope
determination described before can be applied, eg the
ITC-Zorn template method. But one should take special
care to ensure that the true dip is measured. Thus in
a difficult case where the dipping bed does not form a
dip slope, the true dip (θ) can be obtained by the
formula:

$$\tan \theta = \frac{\tan \alpha}{\tan \beta}$$

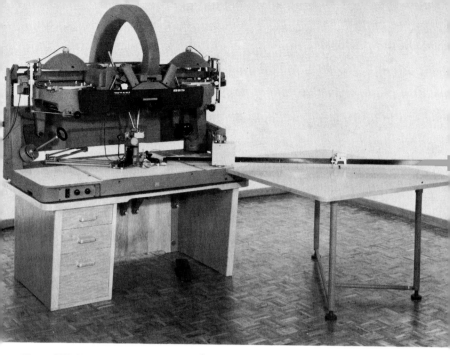

Plate 9 Wild B8S Aviograph *(Wild Heerbrugg Ltd)*

Plate 10 Kern PG-2 stereoplotter *(Kern and Co Ltd)*

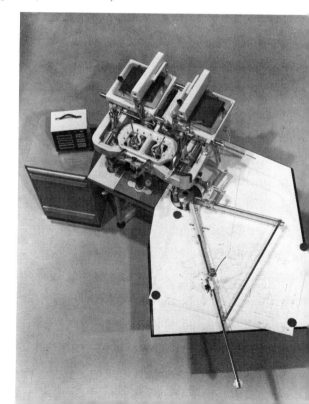

Plate 11 CP-1 plotter *(Cartographic Engineering Ltd)*

Plate 12 PSK-2 Stereocomparator with ECOMAT-21 electronic coordinate recording unit, IBM073 automatic typewriter and Facit 4070 tape punch *(Carl Zeiss, Oberkochen)*

where α is the apparent dip
and β is the angle between the apparent dip direction and
the true dip direction (Fig 3.8).

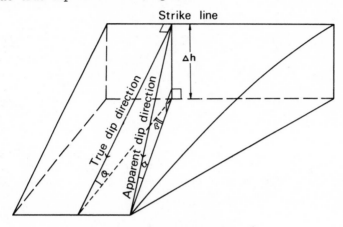

Fig 3.8 *Block diagram showing the*
difference between true dip and
apparent dip

All the methods described above in slope and dip determination are photogrammetric methods which can give rise
to the following types of error: (i) the error in elevation caused by the use of the parallax bar or wedge for
which a standard error of ±0.02mm per point is expected,
so that a standard error of ±0.03mm will result for the
parallax difference;[16] (ii) the error on horizontal
distance caused by relief displacement inherent in the
aerial photograph, which depends on the position of the
points from the principal point and the alignment of the
slope line with respect to the nadir point (principal
point) on the photograph, thus giving rise to two cases:
(a) if the line of strike for the slope passes through
the principal point, the amount of relief displacement
will be negligible (*ee'* in Fig 3.9a) so that the image
distance can be taken for the horizontal distance (the
correct distance should be *de'*); and (b) if the direction
of slope is in coincidence with the direction of displacement (Fig 3.9b) (ie passing through the nadir point),
then the effect of radial displacement will be at its

97

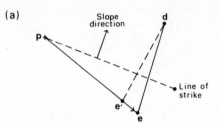

(a)

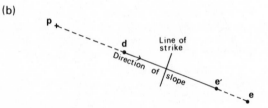

(b)

p = principal point

d = image point for object D

e′ = correct image point for object E

e = displaced image point for object E

Fig 3.9 *Relationship between slope direction and the direction of relief displacement: (a) slope direction at right angles to the direction of relief displacement, and (b) slope direction coinciding with the direction of relief displacement (after Verstappen, 1963)*

maximum, ie the distance *de'* becomes lengthened as *de*, and the amount of relief displacement cannot be ignored. One can conclude from this examination of type (i) and type (ii) errors that the slope measurement can be more precisely done for points selected well apart along the line of slope with a constant gradient.[17]

The same photogrammetric approach can be applied in determining the thickness of a rock formation (ie the bed) and the amount of fault displacements. For the formational thickness, the formula varies according to the nature of the terrain and also whether the beds are steeply or gently dipping. If the beds are steeply dipping, the thickness (t) is given by $t = d/\sin\theta$ where d is horizontal distance and θ the dip of bed (with reference to Fig 3.10a), in which the height of the two points $(A$ and $B)$ measured (ie d) should be about the same. On the other hand, if the beds are moderately steep to gently dipping, the formula for the thickness will become

$$t = \frac{\Delta h}{\cos\theta} + \frac{d}{\sin\theta}$$

where Δh is the height difference (with reference to Fig 3.10b).[18] It is necessary to correct for relief displacement in computing the horizontal distance (d). Erosion that may have occurred in the upper edge of the bed should be compensated for during measurement and the adoption of a floating line as suggested by Desjardins may be useful in defining the upper (or the lower) limit of the bed,[19] and the formula can be varied accordingly as $t = \Delta h/\cos\theta$ (with reference to Fig 3.10b).

For determining the amount of vertical displacement on some faults, parallax measurements can be made with the parallax bar on *marker beds* (ie beds of distinctive lithology) on opposite sides of the fault. The parallax difference gives the amount of vertical displacement caused by faulting.

Quantitative Photographic Data

There are also other types of data extractable from aerial photographs without using any photogrammetric

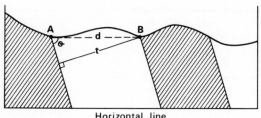

(a)

Horizontal line

(b)

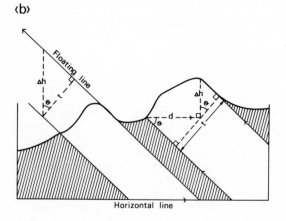

Horizontal line

Fig 3.10 *(a) Steeply dipping beds in comparatively flat terrain (b) Moderately steep to gently dipping beds and Desjardine's floating line method (after Verstappen, 1963)*

machines, such as spectrophotometric and densitometric data which collectively may be called *quantitative photographic* data, according to Ray and Fischer.[20]

1 *Spectrophotometric data* are obtained from measurement of the spectral reflectance of objects imaged on the photograph. This really measures the tonal difference existing between rock types. Spectral reflectance curves can be drawn and it is found that strong tonal differences do exist for different rock types at certain wavelengths as shown for four types of rock samples: (*a*) light-brown sandstone, (*b*) grey limestone, (*c*) red shaly siltstone and (*d*) grey sandstone. Thus, it would be possible to design film-filter systems that would permit easy differentiation of rock types (Fig 3.11).

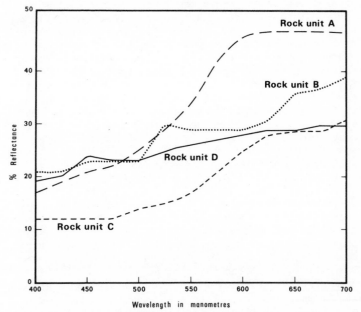

Fig 3.11 *Spectral reflectance curves of fresh samples of light-brown sandstone (A), grey limestone (B), red shaly siltstone (C), and grey sandstone (D) (after Ray and Fischer, 1960)*

2 *Densitometric data* are obtained from measurement of aerial photographs with a densitometer. They give a continuous record of photographic tonal change across the photograph which may be *either* in terms of *optical density* of film materials *or* in terms of *light reflectance* from paper prints. Frequency of tone changes and the magnitude of the tone measurements provide useful descriptions of the terrain features and permit comparison to be made (Fig 3.12).

Quantitative Geomorphology: Network Analysis

In geomorphology, the recent quantitative trend necessitates a stringent re-examination of the underlying concepts of past geomorphic theories such as the Davisian cycle of landscape development, and the hydro-physical approach to quantitative morphology advocated by Horton has been given special attention.[21] Also, the work by Strahler on landform quantification has given further advance in this direction.[22] The important development has been in the concepts of hierarchical organisation of streams, because it has been argued that different segments of a stream system exhibit different morphometric and hydrologic features and relationships. The 'order' of stream is, according to Horton, the numerical group into which it falls in a classification system that defines the smallest unbranched tributaries of the first order. An ordinal scale has been employed by Horton and Strahler for this purpose, and only recently has the ratio scale been employed to indicate order magnitudes as advocated by Scheidegger and Woldenberg.[23] Other quantitative measures of streams advocated by Horton include *drainage density* (ie the average length of stream per unit of drainage area) and *bifurcation ratio* (ie the ratio of the average number of branchings or bifurcations of streams of a given order to that of streams of the next lower order). The stream length can also be measured to compute a stream length ratio with reference to the area of the drainage basin. All these measures can be acquired more easily and accurately from small-scale vertical aerial photographs as illustrated by Shaw, who studied the Lake Nyasa basin in Tanganyika with a plot from 1:50,000 scale vertical aerial photographs.[24] Obviously, the aerial photograph can give a more complete picture of the whole stream system than the

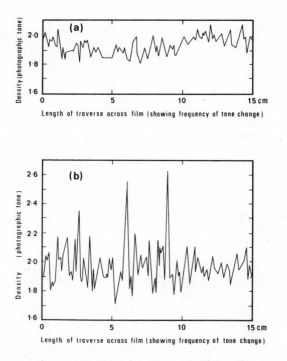

Fig 3.12 *Graphs showing relation of magnitude and frequency of tone changes across two different landforms: (a) alluvial area of low relief, and (b) glacial moraine (after Ray and Fischer, 1960)*

map, ie to maintain the correct proportionality of the network. Recent research has indicated that the 1:10,000 or 1:15,000 scales black-and-white aerial photography provide a suitable degree of generalisation for the measurement of these types of fluvial parameters over areas where extensive tree cover is absent, and computer programs have also been developed employing coordinate

data to describe the network geometry as well as topo-
logic data to describe the network linkages.[25]

*Measurement and Mapping of Morphological Changes in
Geomorphology*

The application of aerial photography is particularly
suited to recording changes of geomorphological features
over time. This involves the use of sequential or time-
lapse photography which provides the basis for comparison
through quantitative measurements as well as mapping.
As some of the changes or movements may be quite small, a
high degree of accuracy must be achieved in the measured
data or in mapping, for which the use of more sophisticated photogrammetric plotting machines, usually in the
class of topographic plotters such as the Wild B8, Kern
PG2 and Galileo-Santoni Stereosimplex IIc, are required.
The provision of good ground controls is also essential.

There are three areas in which such applications of
aerial photography are particularly successful and worthy
of more detailed examination: (1) glaciers, (2) coastal
features and (3) desert dunes.

1 *Glaciers*
Glacial geomorphology has in particular received treatment in this way by many workers such as Finsterwalder,[26]
Case,[27] and Konecny.[28] More prominently, the Department
of Geography in the University of Glasgow has contributed
in this field by following a programme of glacier mapping
in Alaska and Iceland as a cooperative undertaking
amongst photogrammetrists and glacial geomorphologists.
The result is a series of maps showing sequential changes
of the various glacial landform elements such as eskers,
kames, lakes, ice margins, etc, through time, supplemented by photogrammetric measurements of these features.
In a study of the ice wastage and morphological changes
near the Casement Glacier in Alaska, Petrie and Price
have adequately demonstrated the usefulness of such an
approach and shown that the lack of proper ground control
caused most of the troubles and restricted the accuracy
of contour lines that could be plotted from the aerial
photographs.[29] Fortunately, a good knowledge of the
region by the geomorphologist concerning the tidal range
of the sea inlet in the area made possible the use of the

water surface as a datum for absolute orientation. The
other planimetric controls could be provided easily by
stereotemplate triangulation. The kind of aerial photography obtained by means of a Metrogon-lens camera gave
rise to very large radial distortion (±120μm) which had
to be compensated for by means of correction plates when
plotted with the Wild B-8 Aviograph, and the choice of
contour interval (50 ft or 15.2m) in relation to the
original photo-scale (1:39,000), the plotting scale
(1:20,000) and the final map scale (1:30,000) was a
result of the careful consideration of the accuracy
achievable with the plotter (1m in spot heights or about
0.2 per mille of the flying height in this case).
However, checking revealed that the resulting stereomodel
only gave a standard error of ±0.6-0.9m in spot heights
and that the contours of the two plots for 1948 and 1963
exhibited good agreement, making for confidence that any
differences between the two plots were really caused by
morphological changes. All these procedures stressed the
need to maintain a very high metric accuracy if comparison was to be made of the same feature between different
time periods. One should note, however, that for this
type of comparison between two or more maps of glaciers,
ground control requirements are in fact not too critical
provided that all the maps are based on the same control.
From these comparisons, it is possible to compute the
down wastage and volume change of the glacier according
to Finsterwalder's formulae:

$$dh = \frac{dA_1 + dA_2}{A_1 + A_2} \cdot \Delta H \qquad (3.1)$$

where dh is the decrease in height of the glacier surface
during the observation period for the contour interval
$(H_2 - H_1) = \Delta H$; A_1 is the old area, A_2 is the new area;
dA_1 is the difference in area for contour H_1; and dA_2
is the difference in area for contour H_2 (see Fig 3.13);

$$dV = \frac{dA_1 + dA_2 + \sqrt{dA_1 - dA_2}}{3} \cdot \Delta H \qquad (3.2)$$

where dV is the change in volume for the contour interval
$(H_2 - H_1) = \Delta H$;
and $$\overline{dh} = dh/n \qquad (3.3)$$

105

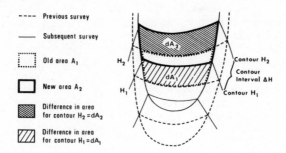
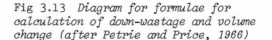

Fig 3.13 *Diagram for formulae for calculation of down-wastage and volume change (after Petrie and Price, 1966)*

where $\overline{dh}$ is the decrease in height for one year, and n is the number of years between the two surveys. This approach succeeded in extracting valuable data on the extent of glacier retreat, decrease in height and loss of volume, from which the rate of modification of landforms in a glacial environment could be computed, thus giving insight into the mode of formation of particular features and leading to the conclusion that the morphological changes in the deglacierised area in front of the Casement Glacier were mainly caused by the changes in the positions of the meltwater channels.

In a follow-up effort of mapping the Breidamerkur Glacier in Iceland undertaken by Welch and Howarth, the same approach has been improved upon to allow more detailed mapping (using the Stereosimplex IIc plotter) for four different years (1945, 1960, 1961 and 1965) and a more comprehensive study of changes in eskers, kame and kettle areas, an ice-dammed lake and coastal features.[30] Examples of such maps showing the development of eskers in the area for the years 1945 and 1965 are shown in Fig 3.14. These may be compared with the aerial photograph of a part of the area (Plate 14). Even panchromatic photography (in this case, false-colour infra-red)

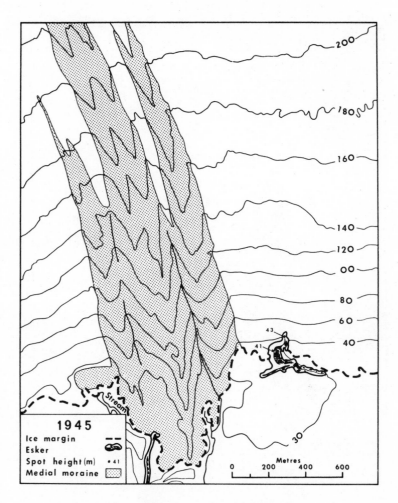

Fig 3.14a *Eskers of the Breidamerkur*
area, Iceland, in 1945, mapped from
aerial photographs

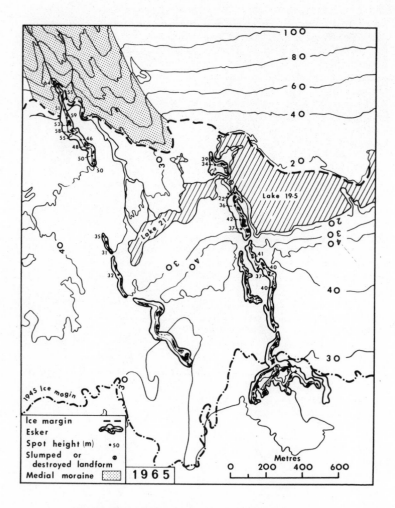

Fig 3.14b *Eskers of the same area in 1965. Note that the lake levels have dropped and esker development continues beneath the medial moraine (after Welch and Howarth, 1968)*

was also employed to study coastal changes as a further
step of research, about which there will be more discussion in Chapter V. All these accurate metric data serve
as good indications of the modes of formation or
destruction of glacial landforms in the region under
study.

2 *Coastal Features and Shore Processes*

Coastal features which also exhibit rapid changes over
time can of course be treated using the same photogrammetric approach. This can provide valuable metric
data on the rate of deposition of such features as
beaches, spits, bars, tombolos, etc, which characterise
a prograding coast; and similarly on the rate of erosion
of such features as the sea caves, arches, sea cliffs,
etc, which characterise a retrograding coast. Such an
approach is best demonstrated in a paper by El-Ashry and
Wanless.[31] By using sequential aerial photographs, long-
term trends in shoreline changes and the development of
shoreline features have been revealed. El-Ashry and
Wanless have shown that from aerial photographs the width
and positions of inlets can be measured; the amount of
sediment eroded or added to the beach can be estimated;
and the amount of sediment carried by longshore currents
or deposited in the form of shoals or tidal deltas can be
estimated by producing with the stereoplotter a series of
contour maps of the shoals at different dates. Thus,
they have been able to show that hurricanes cause the
greatest changes in coastal form through high winds,
waves and tides, but in calm weather after a major storm,
these storm-induced changes are gradually smoothed and
obliterated. They have also ascertained the rate of
cutting back of a spit in Cape Lookout, North Carolina to
be 36.6 metres per year during the period 1945-62. All
these data are essential for a classification of the
coasts, as demonstrated by the work of Alexander in
Tanganyika.[32] Jones has also demonstrated the possibility of mapping the inter-tidal area at one instant
of time for more intensive geomorphic investigation.[33]
The problem of photogrammetric mapping was considerably
simplified in this case since the area was flat so that
each photograph could be treated as a map after rectification and a controlled mosaic could easily be constructed. On the other hand, the establishment of ground
control within the inter-tidal zone has proved to be a

most difficult task because of its rather muddy and marshy nature, as discussed by Hubbard.[34] The placing of ground-to-air markers within the zone is equally troublesome.

One other advantage of aerial photography is to freeze such transitional features as wave and current patterns[35] and other shore processes, all of which can be subjected to photogrammetric measurement.[36] Thus, Cameron demonstrated the use of time-lapse aerial photography and the parallax method to determine the water current velocities offshore.[37] This made use of the anomalous or false stereo effects of the water surface configuration created by eddies, foam lines, etc, which persist long enough to be photographed a second time. When the two similar images are fused stereoscopically, the single image is in a false or anomalous position relative to some stationary objects, such as the adjacent land masses or ships, etc. Such anomalous image can appear to be elevated above or below the adjacent stationary objects respectively, depending on whether the water movements are in the opposite direction to the flight or in the same direction. A simple parallax bar or an elaborate stereoplotter equipped with the floating dot can be used to measure the parallax of this false single image with reference to the nearby stationary object. Since the time interval between the two photographs is known, the speed of the water movement can be calculated. It has been proved that this method of determination of the water velocity has achieved great accuracy and that with standard photography at survey scales ranging from 1:6,000 to 1:60,000, water velocities ranging from 0.4km to 22.5km per hour can be measured. In the case where the water velocities are very high, the displacement will be too great and no false stereo-image can be formed. It is possible, of course, to fly at a higher speed to reduce the time-lapse and obtain the stereomodel. The stereoplotter can also be employed to plot the water current patterns along a river or a coast, provided that a sufficient number of natural or artificial markers are present to give the false stereoscopic effects at various parts of the river or coast, as demonstrated by Forrester[38] who measured both the X- and Y-parallaxes of the image and computed the velocity component in the X-direction (V_x) with the formula

110

$$V_x = z \cdot \frac{B}{H} \cdot \frac{S}{t}$$

and in the Y-direction (V_y) by

$$V_y = \frac{by \cdot S}{t}$$

where B is the length of air base; H is the altitude in the same units as B; S is the appropriate scale factor; t is the time between exposures; z is the elevation of the image obtained from the measured X-parallax and by is the Y-parallax. The same approach with improvements was taken up by the US Coast and Geodetic Survey to produce surface current charts at harbour and inlet areas, but it was also noted that photogrammetric measurements should be made in conjunction with current meter readings for proper reduction of data.[39]

A recent report by Komarov has shown that in the USSR similar interest has been shown in applying aerial photography to oceanological work such as the study of wave disturbances, the mapping of currents and the spectral brightness of the sea, the investigation of other hydro-optical characteristics of the sea and the study of hydrodynamic processes in the coastal strip.[40] It should be noted that the emphasis has been on the use of the spectrophotometer to measure the spectral brightness coefficients of the sea in comparison with a standard. An interesting application where stereophotogrammetry is involved is the measurement of the depth of the sea bed in shallow water areas from large-scale aerial photographs which requires only a slight modification of the parallax formula to the form:

$$\Delta h = i \cdot \frac{H}{P} \cdot \Delta P$$

where i is the coefficient of refraction of the water at a certain position, thus taking account of the fact that two media (air and water) have been involved in viewing the stereomodel.

111

3 Desert Dunes

Features in the desert area exhibit a high degree of instability and change, for the study of which the aerial photogrammetric technique is well suited. One of these features is the sand dunes formed by wind deposition. They are usually found extensively in a flat area and are advancing all the time. A common type of sand dune is the barchan dune which is a crescent-shaped formation with a paraboloid plan and also characterised by having sharp cusps or horns pointing in the direction of the prevailing wind. It has a convex windward slope and a concave leeward slope. There are also other types of sand dune such as the parabolic dune which has a concave windward slope and convex leeward slope in contrast to the barchan; the longitudinal dune which is aligned parallel to the prevailing wind direction; and the transverse dune which lies across the prevailing wind direction. These different types of sand dune can change from one type to the other according to relief, vegetation cover, the strength of the wind, and the amount of sand available. Finkel has illustrated an approach whereby sequential aerial photography can be used to measure the movement of barchan dunes and to examine their pattern of distribution in the Pampa de la Joya area in the Southern Coast of Peru.[41] According to Bagnold,[42] the rate of forward displacement of barchans is a function of the grain size and density of the sand, the velocity of the wind and the dimensions of the barchans. With the use of sequential aerial photography and photogrammetry, this rate can be determined much more easily. Finkel employed two different scales of photograph for two time periods, ie 1:60,000 for 1955 and 1:20,000 for 1958. It is important to have ground controls established and marked, which are also identifiable on the photographs. The Multiplex was used for plotting two maps of over 100 barchans on a dimensionally stable plastic drafting sheet at the scale of 1:10,000. These two sheets were then superimposed and the change in position for each of the barchans was measured with a scale magnifier, thus giving the rate of advance of each barchan. It was found that the barchans advanced at varying rates, with the smaller ones having a greater forward displacement. At the same time, the width (W) of the barchan (ie the distance between two horns at right angles to the wind direction) was measured; and the height of the barchan (H) was also obtained on the ground. It was found that a linear relationship existed between the two expressable by the formula: $W = 10.3 H + 4.0$. The height of the barchan can in fact be measured by means of the parallax method at a lower accuracy from small-scale aerial photographs. The forward speed of the barchans so determined was found to compare favourably with the theoretical speed according to the formula given by Bagnold.

The mass distribution of barchans can also be subjected to a modified quadrat analysis using rectangles aligned with their long sides in the direction of the predominant wind. The photo-mosaic was used for this purpose. The number of dunes falling within each rectangle was counted and the density computed. A statistical analysis was then carried out on a resultant matrix of densities of barchans, from which it was found by Finkel that barchans formed a focus of high density at the south or the upward end of the area studied, from which the average density decreased in the distance down-wind, whilst the density also decreased in the cross-wind direction. It was surprising to discover that the barchans were drifting *en masse* towards the east or up the slope and that the asymmetrical development causing distortion in the barchan shape was due to the sloping of the ground.

From these, Finkel further suggested the use of barchan distribution for geologic dating, having known the rate of advance of the barchans of varying crest heights, the volume of dune material at the source area and the carrying capacity of the wind for this size of sand grain with the assumption that the regime of wind velocities has remained essentially the same through ages. The volume of the dune material at the source can obviously be measured stereoscopically from aerial photographs.

Another example of the application of aerial photographs in studying desert dunes is provided by Clos-Arceduc,[43] who discovered that in the Sahara newly formed live dunes run in a different direction from that of the wind, contrary to the usually accepted theory. It is further suggested after an examination of aerial photographs that only in regions where the dunes are fixed in place by vegetation can aeolian erosion give rise to longitudinal ridges following the direction of the wind.

113

Meteorology

Applications of photogrammetric techniques to provide
metric data of weather elements are also possible, as
demonstrated by the works of Orville, Boge and Pietschner,
all of which involved the use of ground-based stereo-
photography. Orville indicated the suitability of photographs in recording the fast-changing cauliflower-like
cumulus clouds which may herald fair weather on one
occasion and thundery downpours on another.[44] The
photographs were obtained with two metric aerial cameras
set up on the ground, and two exposures for the stereopair were obtained at the same time. The stereomodel was
measured with a modified radial line plotter by which the
tracing of the outline of the cloud was done. Measurement of the range and height of the cumulus clouds was
also made and accuracies of better than 1 per cent in the
6,096m heights and 16.1 - 32.3km perpendicular ranges
could be achieved. The data could be utilised to compute
the vertical velocities and other growth rates of the
clouds.

Boge measured the other element - wind in the upper
atmosphere - using time-sequenced stereophotographs
obtained by two synchronised photo-theodolites set up on
the ground.[45] The vapour trails of the high-altitude
rockets after take-off were photographed at short
successive time intervals, and the coordinate values (ie
X, Y, and Z) of any point on the trail were measured.
The latter measurement can be carried out by using an
analogue universal machine which is capable of taking
large-tilt photographs, such as the Zeiss Stereoplanigraph C8 or the Wild A7. But more conveniently, with the
availability of the electronic computer an analytical
approach was actually used, for which a mono- or
stereocomparator was only required to give the photo-
coordinates. Other corrections necessary, such as the
effects of the earth's curvature, refraction, lens distortion, etc, can easily be incorporated into the
computer program to give more accurate results. From
these results the amount and direction of displacement
of a particular point are known, and given the time
interval between the change in position of the same point,
the wind velocity can be computed from: $v = s/t$ where v
is the wind velocity, s is the displacement of the point

of the vapour trail along any of the coordinate axes in
the time interval t. An accuracy of ±5mph could be
achieved. The great advantage of the method is that
continuous profiles of the velocities of air currents in
the upper atmosphere can be drawn. More recently,
Pietschner described a similar approach in determining
the wind vectors in the lower atmosphere.[46] Near-
vertical smoke trails in red colour which are generated
by special smoke bombs shot up to about 250m using small
rockets were photographed by two synchronised terrestrial
cameras on the ground at intervals of 3 to 5 seconds over
a period of 1 to 2 minutes. The stereopairs were
measured with a Zeiss (Jena) Stecometer Precision
Stereocomparator for the coordinates and X-parallaxes of
those points marking the smoke trails. These data were
recorded on punched tape for numerical calculation by an
off-line procedure with the computer. Difficulties in
obtaining good stereoscopic fusion and in coordinate
measurement were experienced because the smoke trails
showed only few details with little or no extension in
three dimensions. But an accuracy of ±0.03mm for the
X-parallax measurement was achievable for black-and-white
photography. It was found that with the use of colour
photography the accuracy of the X-parallax measurement
would be greatly increased to ±0.015mm. The coordinate
differences of corresponding points appearing on successive pairs of photographs were obtained, from which the
vector components were determined. The analytical
approach was found to be particularly suitable for this
type of application where a large number of photo-
coordinates needs to be processed. This interest in wind
vectors and wind speed measurement has been growing in
recent years as these data contribute towards research
into the diffusion and turbulence of gases in meteorology,
thus helping to minimise the air pollution problem of the
human environment.

Plant Ecology

The other component of the terrestrial environment is the
cover of vegetation on the earth's surface which provides
subjects of study for plant ecologists. Here the use of
aerial photographs has been found to be particularly
suitable. This is because plant ecology deals with the
spatial distribution of plants in relation to environ-

mental factors which may be local or regional. The advocation by Tansley of a holistic appraoch to the subject especially demands an understanding of the interplay of factors and plant distributions for areas of various sizes.[47] Howard believed that aerial photographs are most valuable for ecology study at the macroscopic level, and he introduced the term 'photo-community' to refer to the smallest distinct assemblages of plant species discernible on stereopairs of aerial photographs at a specified scale, ie a phytoscociological grouping.[48] Such a photo-community will vary with the scale and a host of other factors affecting the photograph which has been mentioned in Chapter II. Hence the measurements on the various parameters of the community will refer to different units, which may be *plant formation, sub-formation, association, stand* or even an *individual* plants as the scale of the photograph increases.

The unit of vegetation usually employed by ecologists is the 'stand' which is ecologically defined to be 'any area, the vegetation of which has been treated as a unit for purposes of description',[49] but this is also a term favoured by foresters who define it as 'an aggregation of trees having some unifying characteristics, which occupies a specific area of land'.[50] In general, a stand is taken as a homogeneous unit from the plant-ecological point of view.

There is a need to identify and classify the vegetation found on the earth's surface, and direct measurements from aerial photographs of an appropriate scale can yield parameters for a quantitative description of this vegetation. Exact measurements of the following independent attributes are usually required: (a) *density* or number of plants per unit area, which gives a measure of *relative abundance*; (b) *cover* or the percentage of the ground surface occupied by the vertical projection on to it of the aerial parts of the species of plants under consideration, which gives a measure of *dominance*; and (c) *yield* or the weight (or volume) per unit area per unit time, which gives a measure of *productivity*. More specifically in forestry, these measurements relate to trees and stands such as (1) tree height and stand height; (2) crown diameter; (3) crown closure (or crown density); (4) number of trees; (5) area of stand; and (6) tree and stand volumes; all of which can be directly measured from aerial photographs.

1 *Tree Height and Stand Height*.   Tree heights are best
obtained from stereoscopic pairs of aerial photographs
using the parallax bar and parallax formula.   As has
already been shown in Chapter II, this method can take
account of the various terrain conditions, and, if
required, a statistical adjustment procedure can be
carried out to obtain more accurate absolute height
values.   If only single photographs are available, a
number of tree heights can be obtained by the *shadow
length* method and the *relief displacement* method.   The
relief displacement method makes use of the relief
displacement formula already given in Chapter II.   The
shadow length method involves measurement of the length
of the shadow cast by a tree on the ground.   It is
necessary to know the angle of the sun to the zenith at
the time of photography, which can be found from a nautical almanac or the *Star Almanac* which gives the sun's
declination (δ) at a certain latitude (ϕ) on a specified
day and time.   This in effect gives the sun's angle of
elevation (α) (Fig 3.15).   A straightforward formula can
be applied:

$$h = l.\tan \alpha.H/f$$

where h is the height of the trees; l is the length of
the shadow, α is the angle of the sun's elevation if the
terrain on which the tree stands is flat, and H/f is the
photo scale factor related respectively to the flying
height and the camera focal length.   For non-flat terrain,
a correction needs to be made to the shadow length as
measured directly from the photograph which may be too
long or too short depending on the magnitude of the slope
angle (θ) and whether the terrain is sloping away or
towards the sun respectively as shown in Fig 3.15.
Recently, a review of the shadow height method has been
given by Parry and Gold who also devised a solar-altitude
nomogram to facilitate the solution for the solar altitude
(α), having known the date and time of photography as
well as the latitude and longitude of the location.[51]

The relief displacement method and shadow length method
can be applied only in cases of very *open forest* and
hence they are not suitable for use in tropical and

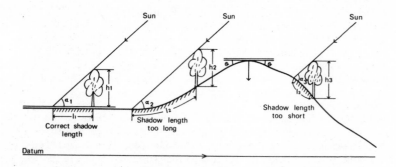

Fig 3.15 *The effect of slope angle and
sun's angle on shadow length*

temperate rain forests.

The parallax method is more generally adopted for tree
height measurement because it can give consistently more
accurate results in many different situations. The
original formula used is

$$\Delta h = \frac{\Delta P \cdot (H_R - h_R)}{P_R + \Delta P} \fallingdotseq \frac{\Delta P \cdot H_R}{P_R + \Delta P}$$

where Δh is the elevation of the point above reference
plane R, H_R is the flying height above the reference
plane R, P_R is the parallax of the reference plane, and
ΔP is the parallax difference between the point and the
reference plane R. This applies when the reference plane
passes through the foot of the tree, which is usually the
case in flat terrain. Without going into elaborate
adjustment, the formula has to be modified in cases of

118

non-flat terrain to take account of the difference in
elevation between the foot of the tree and the reference
plane. Thus, the formula becomes:

$$\Delta h = \frac{H_R \, P_R}{(P_F)^2} \cdot \Delta P$$

or

$$\Delta h = \frac{(H_F)^2}{H_R \, P_R} \cdot \Delta P$$

where P_F is the parallax of the new reference plane F
where the foot of the tree in question passes, and H_F is
the flying height above the new reference plane F.

Analysis of the tree height results obtained in this
way has revealed certain systematic errors caused by
backlash of the micrometer screw in the parallax bar, the
presence of low vegetation (eg a bush) in small gaps in
the forest mistaken as the ground surface, and blurred
tree tops, especially tall trees. The presence of wind
at the time of photography gives rise to Y-parallax or
X-parallax depending on the direction of the wind in
relation to the flight line. The former causes
difficulty in measurement, the latter gives tree heights
too high or too low. It is interesting to note that
paper prints give more error than diapositives, and that
optical enlargement of 3X, 6X, 8X as well as using
photographically enlarged prints do not give a different
result.

Random errors also arise owing to the setting of the
floating mark on the ground and on the treetop. Generally, the accuracy is greater for the terrain than for
the branched treetop as the former is more sharply defined than the latter.[52] It was also found that there is
a significant difference in heighting accuracy between
various operators, according to experience and visual
acuity. In general, to eliminate random errors, it is
advisable to measure more trees for the calculation of
the average rather than to repeat the measurement on the
top of the same tree. The total error in tree height
measurement was found to be ±1.4m for photographs at the
scale of 1:10,000; and ±1.7m for photographs at the scale

119

of 1:20,000, obtained by wide-angle aerial camera (f = 152mm).[53] One very useful point to note is that the tree heights obtained with simple instruments (ie a mirror stereoscope and a parallax bar) compare favourably in accuracy with those obtained with more sophisticated stereoplotters such as the Wild A8; but of course the use of simple instruments is not particularly suitable if a large number of trees are to be measured in this way.

So far, the heighting method of individual trees has been discussed, but usually in forestry only a knowledge of the *stand height* is required, which can be obtained by measuring the heights of a reasonable number of *either* dominant trees *or* co-dominant trees, or both, and then computing the mean height.

2 *Crown Diameter*. Crown diameter is another useful parameter which can be directly measured from medium- to large-scale aerial photographs by means of a micro-meter wedge or open-circle-type wedge. The micrometer wedge consists of two converging lines calibrated to read the separation ranging from 0.05mm to 0.25mm and is usually combined with the parallax wedge. Fig 3.16 shows an example of the ITC wedge. It is moved over the image of the tree crown until the two lines touch the two opposite edges of the tree crown. The open-circle-type wedge consists of a row of open circles with a regular increase in the length of the diameter. It is moved over the image of the crown until the open circle just matches the size of the crown. More sophisticated instruments such as the Teilchengrössen Analysator TGZ3 of Zeiss (Oberkochen) are also available for use in measuring the crown diameters of trees. This simply makes use of a circular light spot shining on to the diapositive plate. The diameter of the light spot can be varied by means of a diaphragm to cover the tree crown, thus measuring its diameter. The results can be registered on a measuring board by means of a foot pedal and the measured diameters are marked on the plate. The results are then converted back to the actual crown diameters of the trees using the scale of the photograph. In any case, for accurate results, only tree crowns near the principal point of the photograph should be measured.

3 *Crown Closure*. Crown closure is the proportion of

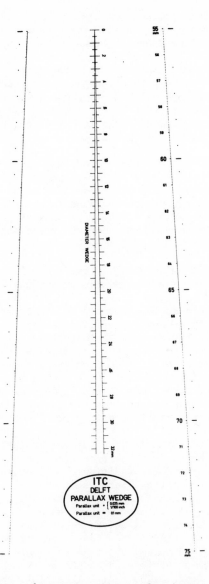

Fig 3.16 *ITC parallax wedge (ITC)*

121

the area of a stand covered by tree crowns. This may
also be expressed as a percentage or a decimal coefficient, which is then known as crown density or crown
cover. Here, the tree crowns are those of the dominant
and co-dominant species only. This parameter is best
measured from a stereopair of aerial photographs viewed
through a stereoscope using a crown density scale. The
scale is really a graded series of dots of varying
densities. Similarly, dot grids or lines can be used for
measurement of crown closure on plots.

4 *Counting of Trees*. Direct counting of the number of
trees or stems can be made on aerial photographs as a
measure of density over an area, but it has been noted
that crown counting from photographs is normally
difficult and inaccurate (usually an underestimation)
since some trees may be hidden or two trees appear as
one.[54]

5 *Area*. Areal measurement is particularly important
in forestry as it is always necessary to determine the
area of a stand or a type class. Direct measurement of
area from photographs is subjected to the error of scale
changes caused by ground elevation differences, as
already noted in the case of geological applications, and
the result should be corrected according to the area
below or above the datum plane from which the flying
height is measured, especially if the amount of relief is
larger than 3 or 4 per cent of the flying height. However, if direct measurement from the photographs is to
be made, it would be advisable to place the photo-plots
at the principal point on each photograph or even on
alternative photographs in order to lessen the amount of
relief displacement. Misinterpretations may also occur
giving rise to biased errors in areal measurement from
photographs. Mathematical methods of correcting for the
combined misinterpretation and scale variation errors
have been developed, notably by Loetsch and Haller[55] and
Stellingwerf and Remeyn,[56] which are for use in conjunction with some sampling devices, such as the dot or
transect method. Usually, the type units are delineated
directly on the contact prints as photo-plots and then
checked in the field. The radial line plotter may then
be used to transfer these photo-plots on to a map, after
which measurement of the plot areas can be made on the

map. It is also possible to make use of a controlled mosaic instead of the map for determining the plot areas.[57]

The usual methods of estimating or measuring the areas of irregular units on maps or photo-mosaics include the planimeter, the transect, and the dot grid. The planimeter, usually of the polar type, requires tracing the boundaries of a certain unit in a clockwise direction (ie measuring the perimeter of the unit by means of a roller, to which the area is related). With careful use, this may yield very accurate results, but it suffers greatly from being tedious and time-consuming.

The transect and the dot grid methods are therefore preferred, and both can be regarded as sampling devices. The *transect* simply involves the use of a series of regularly spaced parallel lines on transparent material to be placed on top of the unit. The line spacing depends on the size of the area to be measured. As shown in Fig 3.17, the shaded area can be obtained as

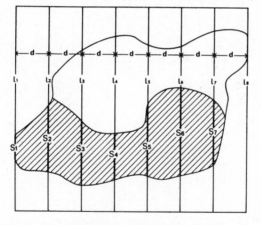

Fig 3.17 *Transect method of finding the area of an irregular shape*

a proportion of the total regular area formed by the parallel lines, ie:

$$\frac{\displaystyle\sum_{i=1}^{n} S_i}{\displaystyle\sum_{i=1}^{n} l_i} \times 100\%$$

where S_i is the length of the portion of the ith parallel
line covered by the unit of area, and l_i is the length of
the ith parallel line of the transect. Since the total
area covered by the n lines transect which is given by
$l.(n.d)$ where d is the spacing between two transect
lines and l is the length of any transect line, the
actual area of the plot can be ascertained from the
previously computed percentage.

The *dot grid* is another sampling device which is more
easily used than the transect.[58] The dot grid, an
example of which is shown in Fig 3.18, consists of

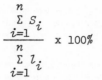

Dot area grid   400 dots—400 cm^2
                                 1 dot/cm^2

Fig 3.18 *The dot grid*

regularly spaced dots within a grid of squares or rectangles on transparent material. The dot density within
each square or rectangle is dependent on the object
under study, the scale of the photograph and the degree
of accuracy required. Wilson suggested the following
formula to estimate the number of dots for a specified
degree of accuracy in large area surveys:

$$N = \frac{P(1-P)t^2}{(E)^2}$$

where N is the number of dots that must be counted, E is
the permissible error expressed as a decimal, P is the
assumed proportion of the total area in the portion for
which an estimate of area is desired, expressed as a
decimal, and t is the value from the t-distribution table
at a specified level of significance.[59] Thus, for a
probability of 95 per cent, t is 1.96 for an infinite
population. When in use, the dot grid template is
placed over the map or the mosaic and the number of dots
falling on each type class is counted, which can be
expressed as a percentage of the total number of dots
enclosed within the grid. As the size of the grid and
the scale of the photograph are known, the area of the
unit under measurement can be computed.

All the above five types of parameter can be directly
measured from aerial photographs. From these, various
other indices of the characteristics of the vegetation
type can be derived. Thus, the *height-crown diameter
ratio* is a reasonably good measure of the *quality* of the
site where the vegetation occurs. In forestry, the site
quality is the capacity of an area to produce tree growth
and obviously is a function of soil, topography and
climate. The *volume of a single tree* can be determined
by correlating between *crown diameter* and/or *tree height*
measured on the photograph and the *volume* measured on the
ground. Similarly, for a *stand volume*, a correlation
between photo-measured stand height and *crown closure*
and terrestrially-measured volume needs to be established.
Alternatively, a correlation between *area* measured on the
photograph and *volume* measured on the ground can also be
applied. From these, an *aerial volume table* can be
constructed for a certain species of tree as illustrated
by Stellingwerf for the *scotch pine* in the Netherlands.[60]

The regression equation therefore assumes the following form:

$$\bar{V} = a + b_1\bar{X}_1 + b_2\bar{X}_2$$

where $\bar{V}$ is the average volume, $\bar{X}_1$ is the average height and $\bar{X}_2$ is the average crown closure obtained from the results of a large number of photo-plots.[61]

There are other similarly obtained indirect measurements from correlations such as (a) the *stem diameter* (or diameter breast high, DBH) which is directly correlated with crown diameter; (b) the *basal area* from crown diameter and tree height measurements, which are employed to describe quantitatively the characteristics of an individual tree.

All these measurements can be made not only on trees but also on other types of vegetation cover on the earth's surface. They provide the criteria necessary for the subsequent classification of the vegetation into *physiognomic types* (if based on the external structure of the vegetation) and *floristic types* (if based on plant species and associations).[62]

Apart from these metric data, the other form of quantitative photography in plant ecology application is the measurement of the spectrophotometric characteristics of individual plant species, vegetation types and genetic soil groups, as has been done also in the case of geological applications discussed earlier.

Urban Morphology

The other component of the terrestrial environment is the artificial environment which bears witness to the great changes brought about by man on the natural environment. This is largely a built environment consisting of houses, buildings, roads, paths and other structures. In other words, it is the shell within which people live, work, and communicate. The layout and form of this collection of building structures result in a townscape which is the subject of study by urban geographers. Such a morphological approach has been advocated by Conzen[63] and Whitehand[64] especially with

the aim of identifying the different cycles in the evolution of the built environment. On the other hand,
Johns adopted an architectural approach to examine the
aesthetic quality of such a built environment.[65] All of
these approaches require the detailed examination of each
individual building for which some quantitative parameters are required. These include the height and the
area of the building, all of which can be directly
extracted from aerial photographs as demonstrated by the
author.[66] The building heights can be obtained by the
usual parallax formula if only simple instruments (a
parallax bar and mirror stereoscope) are involved and
then adjusted using five control points. A mean standard
error of about ±1.14 per mille of the flying height
resulted. It is interesting to note that, as in the case
of tree height measurements, the height for the base of
the building is more accurately measured than that for
the building top. But the accuracy of measuring heights
of buildings from aerial photographs is also affected by
the types and styles of the buildings, as in the case
of tree height accuracy being affected by the tree
species. Building height measurements made by the author
for Glasgow and Hong Kong city centres revealed that the
building heights obtained in the Hong Kong case are more
accurate than those in the Glasgow counterpart, probably
because the Hong Kong buildings are usually high-rise and
more simplistic in form (flat-top) than the more elaborate pitched-roof buildings of Glasgow; the former type
with flat roofs allows more precise setting of the
floating mark than the latter.

As for the measurement of the area of the building,
the same problems already discussed earlier on photo-plot
area measurement in forestry, ie variations in ground
elevation and relief displacement, are encountered. The
dot grid and the transect methods can also be employed
for measuring building areas. An analogy between
buildings and trees has been stressed here with the hope
that the techniques developed for forestry can be
applied to buildings. It also implies that the other
measures for trees are equally applicable to buildings,
such as crown diameter (equivalent to roof-cover area),
crown closure (equivalent to building density), etc.

Population Estimation

Another useful type of metric data obtainable from aerial
photographs for the built environment is the number of
inhabitants. This type of data, of course, is not
strictly photogrammetric in the sense that it can be
directly measured; but at least in one approach, the
frontage of each house has to be measured as a correlate
to the actual population number. This idea has been
applied to the city of Leeds where aerial photographs
at a scale of 1:10,000 were used to yield an estimate
of its population.[67] A qualitative interpretation was
first carried out to identify the three most common types
of house in the city, namely, back-to-back, terraced, and
semi-detached. The strip lengths of back-to-back and
terraced houses were measured to the nearest millimetre
(using the scale magnifier) and the semi-detached houses
were counted. Where the terraced houses had three
storeys the measured length was multiplied by 1.5. The
population figures for each of these enumeration districts were obtained from the census reports and were
related to the amount of housing found on the aerial
photographs. From these a photo factor (PF) was
calculated. On the basis of ten enumeration districts
for each housing type, the average photo factor (APF)
was calculated for each. Thus, by multiplying the
appropriate APF to the strip length of a certain type of
house, an estimate of population can be obtained. It was
found that the overall population estimates obtained in
this way showed an error of only 2 per cent which was
mainly caused by (a) misidentification of housing types
and (b) the universal application of an average photo
factor. Another approach to estimating population from
aerial photographs involves placing a transparent
floating grid over the aerial photograph (usually at a
large scale; in this case, 1:5,000) and counting the
number of houses found within each grid cell by the dot
method.[68] The average household size (ie the average
number of persons per household) is then obtained for
the district, and the population size is found by multiplying the number of persons per household by the number
of houses counted. This implies of course that one house
(or one unit of dwelling) contains one household only.
The accuracy of this method of population estimation
depends on the skill and experience of the photo-

128

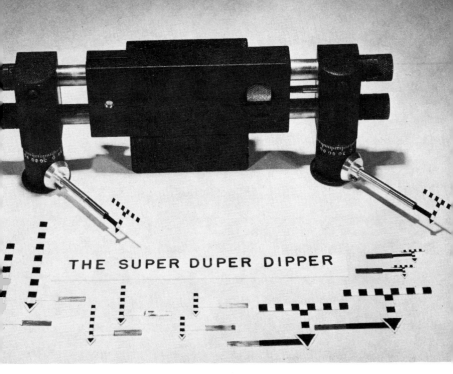

THE SUPER DUPER DIPPER

Plate 13 Stereoslope Comparator *(Robert J. Hackman)*

Plate 14 A stereogram of the Breidamerkurjokull Glacier, Jokulsarlon, Iceland, taken with a Wild RC5 Aviogon camera (focal length = 152.44mm) at an altitude of 4,100m, thus giving a nominal photo scale of 1:27,000. Notice the following features: (A) crevasses on the surface of the glacier, (B) lateral moraine, (C) melt-water lake, (D) the outwash plain, (E) moraine ridges marking the edge of the glacier at different stages of retreat, (F) eskers and (G) kettle holes *(Department of Geography, University of Glasgow)*

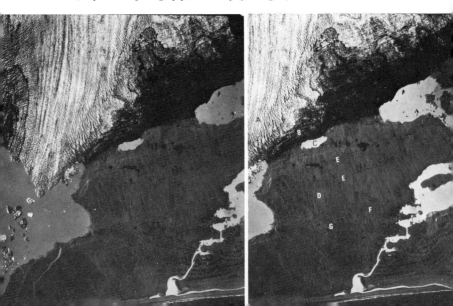

Plate 15 A stereomodel of the front façade of Kelvin Lodge in Park Circus, Glasgow *(Department of Geography, University of Glasgow)*

Plate 16 A stereogram of Shui Tau near Kam Tin in the New Territories of Hong Kong (taken with a Wild RC8 camera with a focal length of 151.96mm at an altitude of 1,189m). Note the tonal variations according to the depth of water in the fishponds (cf A, B, C); the different surfacing materials for the road at D and the path at E; the crop textures and the patterns of the vegetables at F and the fallowed paddy fields at G; and the compact square-shaped pattern of the Chinese settlement at H *(Crown Lands and Survey Office, PWD, Hong Kong Government)*

interpreter in the identification of houses, and in most residential areas the random error was less than 3 per cent. One should note, however, that such a method of population estimation presupposes homogeneity in housing type and uniformity in household size. These do not necessarily occur in other places, especially non-Western cities such as Hong Kong where shared households are common and where housing types are highly heterogeneous and mixed. In general, there is a great contrast in the characteristics of the built environment between the developed and the developing countries as a result of socio-cultural differences. Eyre *et al* have pointed out the usefulness of time-sequence aerial photography in supplementing the census material for population analysis. Examples drawn from Jamaica were given to illustrate how aerial photography increased the validity of the census and detected inadequacies. It was concluded that with the careful use of sampling techniques, aerial photographs can be employed to extrapolate demographic data in areas without detailed census coverage.[69]

Transport Planning and Traffic Studies

This involves measurement of moving objects on aerial photographs. In this field of highway engineering and transport planning, there has been an increasingly acute demand not only for topographic and geologic data of the terrain, but also for accurate and up-to-date traffic flow parameters, such as vehicle speed, position, spacing and volume, which in turn determine the technical standard of the highways or highway systems.[70] The usefulness of photogrammetry in different stages of high-way design has already been confirmed. However, special care needs to be taken in acquiring suitable aerial photography for the purpose in hand, as has been demon-strated by Stoch *et al.*[71] The successful accomplishment of the two essential operations of vehicle identification and vehicle count, required in most traffic surveys, depends largely on the scale of the photographs and the time interval between exposures. In particular is stressed the problem of selecting a suitable scale of photography as a unique one for this kind of application of photogrammetry because the objects involved are totally different from the objects normally dealt with, ie topographic features. On the other hand, even within studies, widely different circumstances exist, for example, in studying moving vehicles in the *traffic* stream or stationary vehicles in parking areas. All these conditions have to be taken into consideration in selecting a suitable type of aircraft, camera, flying height and time interval of exposure for aerial photography. In these respects, two interesting approaches to the obtaining of traffic parameters are worth examining in greater detail.

1 *Determination of traffic flow parameters by the sequential exposure method of photography.* This is an approach experimented with by Wohl and Sickle who employed continuous strip photography with the Sonne camera to give an unbroken and lengthwise coverage of the highway under study - a type of *non-conventional photographic system* already explained in Chapter II.[72] In this application, stereopairs of continuous strip photographs were obtained with a dual-lens cone which displaces the right lens forwards and the left lens backwards to create a stereobase in the direction of flight. The intersection of these forward and backward lines of sight produces the parallax angle. In this way, a stereopair of continuous strip aerial photographs covering a highway for about 145km has been used by Wohl and Sickle with a flying height of about 305m, a focal length of 101.6mm and a parallax angle of 5°. From this, they developed a number of formulae to calculate the speed, spacing and volume of vehicular traffic. Thus, the formula for calculating the speed of vehicles is

$$D = 2H.\tan\tfrac{1}{2}\phi \tag{3.1}$$

where ϕ is the parallax angle, D is the air base and H is the flying height of the aeroplane. Since there is a time-lag between the front and rear pictures, then:

$$D = S_p(t'-t) \tag{3.2}$$

where S_p is the speed of the aeroplane, and $(t'-t)$ is the air base time-lag, and by substitution of 3.1 in 3.2,

$$(t'-t) = (2H \tan\tfrac{1}{2}\phi)/S_p \tag{3.3}$$

This can be applied to a vehicle B in the stereopair
taken by the front lens at time t_1 and rear lens at time
t_2; and as the aeroplane has flown over a distance D_p,
the vehicle B has also moved over a distance D_b, ie,

$$D_b = S_b(t_2-t_1) \qquad (3.4)$$

$$D_p = S_p(t_2-t_1)$$

Therefore, $\qquad D_p = D_b+D \qquad (3.5)$

Thus, the formula for the moving vehicle speed, S_b, is

$$S_b = (S_p.D_b)/(D_b+D) \qquad (3.6)$$

This is clearly a straightforward derivation. Similarly,
Wohl and Sickle also developed a formula for determining
the vehicular volume, V (or the actual number of vehicles
per unit time) as follows:

$$V = \frac{\bar{s}.(n-1).S_p}{D_t.(S_p-S_d)} \qquad (3.7)$$

where $\bar{s}$ = average overall speed of n vehicles, D_t =
distance between end vehicles (as scaled from the photograph), and S_d = speed of the last vehicle in line.

From this experiment, Wohl and Sickle concluded that
continuous strip photography can be further extended to
acquire other types of traffic parameters.

2 *Determination of traffic flow parameters by the
analytical stereoplotter.* Taylor also reported on an
analytical approach to extract accurate traffic flow data
from aerial photographs, namely, vehicular spacing and
velocity, continuously in time and space, which may be
shown graphically as trajectories of all the vehicles
present on the highway at the same time.[73] This simply
involves the direct measurement of the position of each
vehicle with reference to an arbitrary coordinate system
from the photographs, which are then transformed to the
ground coordinate system. Taylor made use of a non-
metric aerial camera, a Maurer P-2 70mm reconnaissance
camera (with Kodak lens, f/2.8, focal length = 76.2mm),

mounted on the helicopter to obtain vertical aerial photographs at an approximate scale of 1:12,000. A test car was also used for accuracy checks and the traffic surrounding it was photographed. A sequence of 101 photographs from one flight was analysed in detail. From the photographs, ground control points adjacent to the highway were selected. These included features like lamp-posts, manhole covers, guard rail posts and ends, paved ramp noses and other roadside features, which could be positively identified on the 1:12,000 scale photographs. The grid coordinates of these control points were established by a third-order ground survey. A total of 41 ground control points was required for these 101 photographs. Such an analytical approach is obviously tedious as a large number of points are involved, and the use of a computer is essential. Normally, a stereocomparator or a monocomparator can be used to measure the photo-coordinates of points precisely. Taylor, however, made use of the more advanced OMI-Bendix AP/C Analytical Stereoplotter to measure the photo-coordinates and the use of the IBM-7094 computer for the transformation. The AP/C Analytical Stereoplotter has the great advantage of possessing a high degree of adaptability to virtually any type of photography which includes narrow-angle, normal-angle, wide-angle, and super-wide-angle; focal lengths from zero to 1,220mm; and vertical, tilted, convergent, oblique or panoramic photography. It also takes into account any type of lens distortion, film shrinkage, atmospheric refraction and earth curvature. Therefore, the errors of the helicopter photographs obtained with the Maurer P-2 reconnaissance camera can easily be corrected. Altogether a total of 3,700 spacings and velocities were computed and the mean errors in the spacing and velocity determination were found to be 0.2m and 0.64km per hour respectively.

Reconstruction of the Human Environment of the Past

Aerial photographs which record the present condition of the terrestrial environment can also be interpreted to give a remnant picture of the past, provided that the angle of sunlight reflection is right to bring out a proper tonal contrast between the old and the new. Even buried sites of which no traces whatsoever are visible to an observer on the ground can be easily discovered from the aerial photograph against this matrix of tonal variations. Thus, it is possible for historical geographers and archaeologists to trace on aerial photographs many phases of man's past and present activities - social, economic and military - and also innumerable aspects of his natural background.[74] The technique of photogrammetry comes in again to help reconstruct ancient human settlements, such as shown by the work of Fairhurst and Petrie.[75] Their interest was in mapping by means of aerial photography the past environment and the deserted Highland settlements called clachans in Scotland. A series of maps at the scale of 1:10,000 were plotted by a Wild A-6 machine to show features such as old roads, dykes, buildings, etc, of the two clachans Lix and Rosal at different time periods. It has been specifically noted that such mapping was possible because of the availability of ground controls - a condition which needs to be satisfied before any applications to utilise the metric aspect of the stereomodel are possible.

An extension of this technique is to map individual buildings or monuments of historical value, which are either still in existence or ruined. The metric data obtained can be used for restoration purposes in case of damage. This involves the use of terrestrial photogrammetry and arouses an interest in so-called *architectural photogrammetry* advocated as early as 1858 by Meydenbauer.[76] From then on, interest in this type of special application grew in Europe where elaborate buildings and monuments of great architectural value such as cathedrals, palaces, castles, etc, abound. Terrestrial photogrammetry, which may be regarded as a special case of aerial photogrammetry, involves only an interchange between the Y- and Z-axis in measuring the stereomodel.[77] An example of application by Thompson has shown how effectively the data could be utilised to restore the building of Castle Howard.[78] As a result, a Royal Commission on Historical Monuments was established in England in 1963 to carry out this type of architectural recording.[79] All applications invariably involved the mapping of the façades of the buildings and the meticulous depiction of all the architectural details that were visible on them. In Fig 3.19 an example is shown of the photogrammetric plot of the front façade of Kelvin Lodge in Park Circus, Glasgow, which should be

Fig 3.19 *The photogrammetric plot of the front façade of Kelvin Lodge in Park Circus, Glasgow (D.A. Tait, University of Glasgow)*

seen together with the photographs from which it was plotted (Plate 15). The topographic plotter Galileo Stereosimplex IIc was used for the plotting.[80] In recent years, interest in this type of application has been flourishing to such an extent that special photographic and plotting instruments have been developed by the major manufacturers, such as Wild and Zeiss.[81] More difficult applications have been attempted, such as in the mapping of domes, and an increasingly analytical approach has been followed.[82] Closely related to this type of application is the mapping of rock faces using terrestrial photography to produce a contour map of the rock surfaces, representing depth measurements beyond a specific vertical reference plane. Both the Edinburgh Castle Rock and the rock face of the Niagara Falls have been mapped in this way.[83] These maps, however, will be of interest to engineering geologists rather than to historical geographers. In all these applications, points have to be marked on the façade of the building or on the monument or on the rock face, which have to be surveyed accurately for use as control points in the absolute orientation of the stereomodel.

CONCLUSIONS

In this chapter, the role of aerial photography in
providing metric data of our terrestrial environment in
both the digital and cartographical forms has been
stressed, and it is quite clear that there are many areas
of interest to geographers where photogrammetry can be
successfully applied. The main advantage of the
technique lies in the fact that only relatively simple
concepts need be involved and it can be applied at
varying levels of sophistication to meet different
requirements of accuracy. It is noteworthy that both the
analogue and the analytical solutions are possible. But
for more universal applications in geography where the
metric aspect is involved, a more sophisticated stereoplotter such as one of those in the topographic plotter
class is particularly suitable as it can speed up the
whole data-acquisition procedure and gives geographers
an efficient tool to explore the spatial relationships
of geographical phenomena which are so effectively
portrayed by aerial photographs in three dimensions.

IV AERIAL PHOTOGRAPHS AS DESCRIPTIVE MODELS

THE QUALITATIVE NATURE OF AERIAL PHOTOGRAPHY

The qualitative characteristics of aerial photography are inseparable from its metric counterparts, as one must know what one is measuring. This means that photogrammetry has to go hand in hand with photo-interpretation, which explains why in topographic mapping photo-interpretation also has an important role to play, as observed by Tait.[1] It is not surprising therefore that most geographical applications have taken full advantage of the fact that aerial photographs are descriptive models of the reality.

One other important fact about aerial photography is that the photographs are capable of being interpreted with specific themes in mind. Thus, they can be interpreted to give physical information only, but the same aerial photographs can be reinterpreted to give distinctly different social information. This characteristic has been specially noted by the French sociologist, Chombert de Lauwe, as being most suitable to illustrate the concept of social space.[2] He regards social space as the total environment made up of a morphological environment and a socio-cultural environment perceived by the human mind. The morphological environment consists of a number of sub-spaces, and from the same set of aerial photographs of the environment, each of these sub-spaces can be separated. De Lauwe's concept is extremely useful and in fact forms the basis of most photo-interpretation work. But it must be noted that all of these sub-spaces are not equal in degree of complexity. Thus, social morphology is more implicit and complex than physical morphology in the urban environment. Similarly the external features of the physical environment are more easily interpreted than the internal socio-economic characteristics of the human environment.

136

THE NATURE OF PHOTO-INTERPRETATION

The formal definition of photo-interpretation given by the American Society of Photogrammetry states that it is 'the act of examining photographic images for the purpose of identifying objects and judging their significance'.[3] A Russian view is that interpretation should be regarded as a definite creative process which establishes the presence of objects and phenomena in the studied area, determining and explaining the causal relationship between them.[4] From these definitions it is quite clear that there are three distinct stages in photo-interpretation: (1) the examination stage, (2) the identification stage and (3) the evaluation stage which can be treated as the classification stage.

Similarly, Vink in a stimulating treatise on photo-interpretation has identified six types of photo-interpretation: (1) detection, (2) recognition and identification, (3) analysis, (4) deduction, (5) classification and (6) idealisation.[5] *Detection* is selectively picking out objects (directly visible) or elements (indirectly visible) from the aerial photographs. These directly visible objects are then *recognised and identified*, to be followed by a process of analysis which delineates groups of objects or of elements possessing individuality. Then the more complicated process of deduction is involved by applying the principle of convergence of evidence to predict the occurrence of certain relationships. *Classification* comes in to arrange the objects and elements identified into an orderly system, and finally through *idealisation* a line is drawn as the ideal or a standardised representation of what is actually seen in the photo image.

Thus, Vink saw photo-interpretation as a series of interconnected processes progressing in steps from simplicity to a high degree of sophistication. He also pointed out the importance of the human interpreter and the *level of reference* which governs the nature of the outputs from photo-interpretation. The level of reference is defined as the amount of knowledge which is stored in the mind of any person or group of persons interpreting photographs.[6] There are different levels of reference - general, local and specific. The general

level is the interpreter's general knowledge of the
phenomena and processes to be interpreted, the local
level is the interpreter's intimacy with his own local
environment which adds extra knowledge, and finally the
specific level is a highly specialised one, involving
a much deeper understanding of the processes and phenomena he wants to interpret.

Within this framework provided by Vink, one can
perhaps simplify the relationship by rearranging these
processes in the form of a flow chart as shown in Fig 4.1.

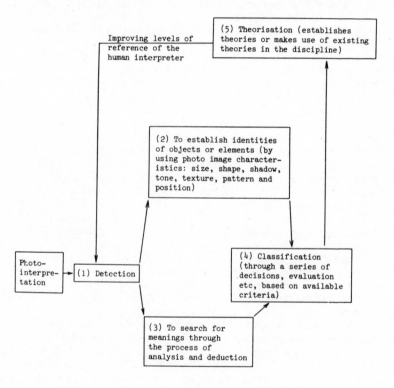

Fig 4.1 *Processes of photo-interpretation*

For many purposes, it suffices to consider photo-interpretation as essentially consisting of two processes: (1) a
process of establishing the identities of the objects
and elements detected in the aerial photographs, and (2)
a process of searching for their meanings. In the first
process, the photo-image characteristics, such as size,
shape, shadow, tone, texture, pattern, and position, are
used to help identify the objects, whilst the more
sophisticated processes of analysis and deduction are
used to discover meaningful relationships in the second
process.   The result is a classification which attempts
to impose some sort of order and coherence on the
qualitative information obtained. Classification leads
to theorisation, through which the interpreter's levels
of reference are raised, thus improving his skill in the
detection of important objects or elements from aerial
photographs, a process which precedes identification
and searching for meanings in photo-interpretation.

Thus, photo-interpretation is basically a classificatory process which aims at assigning photo-images
into their proper groups so that an aggregate pattern
can be brought out or their interrelationships revealed.
Classification involves human decision, so that photo-
interpretation is also essentially subjective.

METHODS OF PHOTO-INTERPRETATION

In the foregoing discussion it has been noted that
photo-interpretation proceeds in stages, and for high
efficiency a systematic approach is preferred. A three-
stage procedure is usually followed:[7]

Stage 1: General Examination. This aims at getting a
general impression of what the area as shown on the
photographs is like. The general patterns of relief,
vegetation, and cultural development are to be established. This really includes Vink's detection phase of
photo-interpretation.

Stage 2: Identification. Concentration on more
specific areas or features is required in order to
identify and recognise the individual objects. This has
to be done in a systematic manner. Normally, the
following two approaches have to be used:

1 *By significant areas.* The most significant areas
established in the general examination stage are examined
in detail one by one. For example, a built-up area may
have been identified in the first stage as the most
significant. In the second stage, the characteristics of
this built-up area are studied to identify street patterns, building types, and the location of some special
structures such as churches or temples, cemeteries,
public parks, etc. After this, the second most significant area is similarly studied in great detail, eg the
agricultural area; and then the third most significant
and so on, finally ending with the least significant.

2 *By uniform sub-areas within a significant area.*
Within a significant area (or if the photograph shows a
uniform type of landscape), a further sub-division into
smaller areas is arbitrarily made, and each such area is
then examined in detail in turn. This facilitates a
systematic coverage of the whole photograph during
interpretation.

The identification and recognition of objects are
helped by a knowledge of the characteristics of photo-
imagery as recorded by black-and-white panchromatic film.
These include tone, texture, pattern, shape, shadow,
size and situation.[8]

(a) *Tone* represents a record of light reflection from
the land surface on to the film. On panchromatic film
it will be seen as ranging from black, through greys,
to white. In general, the more light reflected by the
object, the lighter its tone on the photograph. The
nature of the materials affects the amount of light
reflected. Thus, bare earth usually appears to be light
in tone; but if it gets wet, its tone will become darker.
However, the angle of the sun appears to have an over-
riding importance in determining the tone of an object.
Thus, one may say in general that water surface appears
to have a fairly dark tone on one photograph, but on
the next photograph, it can be white if the sun's rays
are all reflected from it on to the film. Usually, the
steeper the angle at which the sun's rays reach the
surface of an object, the brighter is its illumination.
Consequently, in areas of dissected relief, slopes with
different degrees of steepness and different aspects are

illuminated differently, thus producing different tones on photographs. Therefore, the tone of an object as imaged on the photograph is influenced by many factors and one needs to exercise great care in evaluating it. Plate 16, which shows a fishpond area in the rural New Territories of Hong Kong, illustrates very well the usefulness of tone in determining the depth of water inside fishponds, the degree of sedimentation in the river and the wetness of the soils. Similarly asphalt-surfaced roads and concrete-surfaced paths can be differentiated.

(b) *Texture* is the frequency of tone change within the image which arises when a number of *small* features are viewed together. Obviously, this 'smallness' of the feature is determined by the scale of the photographs used; hence the texture of a forested plot as seen at the photo-scale of 1:50,000 will be different from that as seen at the scale of 1:1,000, as the tree crowns will contribute to the overall textural appearance of the image at the large scale. In general, one can distinguish between the *smooth*, *mat* and *rough* textures as represented by the calm water surface, ploughed field and forested area respectively (Plate 16). Another particularly good example is given in Plate 17 (Kent) from which one can differentiate, using texture alone, the different types of crops - mature small fruit trees (A), hops (B), mature large fruit trees (C), young fruit trees (D) and vegetables (E).

(c) *Pattern* is a familiar characteristic of the landscape to geographers as it results from the spatial arrangement of objects. Normally a macroscopic viewing of the whole photograph is required to detect the significant pattern. Examples include the field patterns, settlement patterns, street patterns, drainage patterns, etc. In Plate 16, the patterns of the paddy and vegetable fields can be compared with those in Plate 17. The compact and regular layouts of the Chinese Villages should also be noted.

(d) *Shape* is a qualitative variable describing the general form, configuration or outline of an object which is particularly useful as a single factor for the identification of objects. But in vertical aerial photographs,

the shape of an object assumes the unfamiliar plan view
from the top. With the addition of the third dimension,
ie height achievable by the creation of a stereomodel,
objects are more easily identified in conjunction with
the shape variable. However, one should note that the
shape of a vertical object will vary according to its
position on the photograph (ie near the principal point
or not). In Plate 18 (Scunthorpe) round objects of
different sizes and heights can be easily detected but
they represent very different things which include cooling
towers (A), gas storage tanks of varying sizes (B, C)
and a concentration tank (D).

(e) *Shadows* of objects usually enhance the usefulness
of the shape factor in identification and recognition,
especially for linear vertical features such as a row
of trees, lamp-posts, telegraph posts, stone walls,
fences, hedges, etc, and also structural weaknesses such
as faults, joints, fracture lines, etc, on the earth's
surface. On the other hand, excessively long shadows
can cause trouble to photo-interpreters and photogrammetrists by obscuring the ground surface and other
features in the paths of the shadows. In Plate 18, all
these points have been very well brought out.

(f) *Size* (the dimensions of an object) needs to be
carefully related to the photo-scale. Usually, the best
approach is to compare an object of unknown size to some
common objects of standard dimensions on the photograph,
eg a football field, a car, etc, if such are available.
Otherwise, the nominal scale of the photograph has to
be noted and a rough measurement made directly to the
object on the photograph for an estimate of its size.

(g) *Situation* involves considering the object's
position in relation to others of known significance in
its immediate vicinity. The method of association is
utilised in this connection to identify features. In
reverse, the presence of some special features implies
the fulfilment of the prerequisite conditions for their
existence. Thus, the positive identification of a
mangrove indicates that the area covered by this
vegetation has a coastal situation and is periodically
flooded by sea water.

142

All these seven characteristics of the imagery serve
as useful aids to photo-interpretation at this identification and recognition stage; and with experience,
their employment becomes automatic.

Stage 3: Classification

This stage includes Vink's analysis, deduction and
classification types of photo-interpretation. As has
already been pointed out, classification involves
assigning photo-images into their proper groups. For
example, roads may be classified into first, second and
third class according to their width and functions;
similarly, a typology can be developed for crops,
buildings, settlements, etc. In this stage, the evaluation of the significance of individual images as a
whole is made, from which *generalisations* (or *idealisation* according to Vink) can be formulated.

APPLICATIONS OF PHOTO-INTERPRETATION
TECHNIQUE IN GEOGRAPHY

Geographers have long been concerned with the use of
aerial photographs as a major source of qualitative data,
and applications of photo-interpretation techniques are
highly varied. Based on the degree of sophistication
with which the technique has been employed, one may
distinguish two major directions of approach: (1) by
direct observation; and (2) by indirect observation.
Both of these have been applied to extract qualitative
data from aerial photographs for the natural environment
as well as the artificial environment.

By Direct Observation

These include all applications using photo-interpretation
to extract qualitative data of those phenomena which are
directly observable on aerial photographs. Thus, parameters describing the *external* characteristics of the
terrestrial environment are obtained. Such applications
can be again sub-divided into two groups according to
whether the natural or the artificial environment is
studied, but in both cases the photo-interpreter
concerned must focus on a particular aspect of interest
(ie with a specific objective in mind) and exercise his

three levels of reference, ie general, local and specific, to arrive at an identification of features as accurately as possible. It should be stressed again that the use of aerial photographs does not dispense with field surveys completely but only minimises the need for them. Photo-interpreted features need to be checked on the ground to establish the degree of accuracy. Hence, a usual approach of photo-interpretation is to develop a key for aiding the identification of objects in some sample areas and then applying it to other previously unsurveyed areas.

A photo-interpretation key has been developed on the principle of identification keys employed by field workers in the biological and physical sciences.[9] The purpose is to help the observer to organise the information present in aerial photographs and guide him to a correct identifcation of unknown objects. The key is a system-atic listing and description of the distinctive charac-teristics of the features to be identified supported by illustrations with annotations, which may present the vertical plan view in a stereopair or as a single photo-graph and/or on oblique or even terrestrial views of the same features.

There are generally two types of photo-interpretation key depending on whether the principle of (1) selection or (2) elimination is employed for their construction. In the first case, the different classes of phenomena are illustrated and described, and the photo-interpreter selects the one which comes closest to the features examined. In the second, a series of possible identi-fications are presented step by step, usually by means of a dichotomous system of paired choices, and the photo-interpreter is required to eliminate the one which is incorrect at each stage until he reaches the final answer. The usefulness of these keys has been the subject of argument among photo-interpreters, but it has been shown by a quantitative evaluation that if the material is reasonably well organised, no significant difference should exist between the two.[10] However, one should note that the dichotomous key is designed to force the inter-preter to work from the general to the particular by leading him through a mass of detail to a final answer. On the other hand, the selective key does not give great detail but treats the subject as a whole. It follows therefore that the dichotomous key is more difficult to prepare than the selective key bacause it has to provide rather detailed information at each stage. In forestry, dichotomous keys are usually employed. An excellent example is given by Sayn-Wittgenstein who made use of crown characteristics to identify tree species.[11] An example of a good selective key is one dealing with industrial land use which was constructed by Chisnell and Cole.[12] There are also other keys prepared for landform studies such as those by Powers,[13] and Waldo and Ireland.[14]

There are also some more specialised types of key such as the regional keys[15] and analogous area keys,[16] both of which make use of the concept of regional geography. The former classifies the terrain into homogeneous units such as physiographic units whilst the latter attempts to develop the key in an accessible area for application to an inaccessible area, having been based on the assumption that every geographic region in the world has at least one analogous counterpart elsewhere. This is really an important approach which represents an early attempt to look for some general laws in regional geography.

In conclusion, one should note that the photo-interpretation key only helps to identify, not to interpret; and the success of such keys must be judged, as Bigelow rightly points out, against the functions that they were originally intended for.[17]

In the following discussion on the various past applications of the direct observation approach in photo-interpretation, two areas of interest to geographers can be distinguished: (1) landform analysis, and (2) land-use analysis.

1 *Landform Analysis*. The external characteristics of the natural environment as imaged on black-and-white panchromatic film provide the impetus for a morphological approach which results in a description of the different landform types on a regional basis. Thus, photo-interpretation has been found to be particularly effective in providing information on the distribution of features associated with the following types of landform:

(a) *volcanic landforms* such as cones, craters, crater lakes and lava flows;[18]

(b) *karst landforms* such as sinkholes, karst cones, ponors and poljes and the calculation of the 'degree of karsting' from drainage, hole-density and peat development in an area;[19]

(c) *glacial landforms* such as eskers, kames, kettles, ice-dammed lakes, drumlins, moraines and glaciated valley forms;[20]

(d) *landforms resulting from sub-aerial erosion and mass wasting* such as gullies, landslides;[21]

(e) *riverine landforms* such as drainage patterns, forms of river channels, terraces and deltas;[22]

(f) *coastal landforms* such as tidal flats, marshes, beaches, seacliffs, sea caves, arches, stacks, reefs, etc;[23]

(g) *aeolian landforms* such as sand dunes and loess;[24] and

(h) *desert landforms* such as playas, bajadas, pediments, hamada, and inselbergs.[25]

All these landforms produced by aggradational as well as degradational forces can be easily interpreted from aerial photographs by evaluating the image characteristics of the various objects, in particular, tonal contrasts and patterns. Based on these, individual landforms and features are then classified in accordance with the accepted terminology in geomorphology. The latter is a detailed landform classification based on field observations and therefore functions analogously as an interpretation key. As an example, it is not enough to identify a landform as a flood plain on the photograph; it is necessary to indicate at the same time whether it is a meander flood plain, a covered flood plain, a composite flood plain, or a bar meander flood plain, etc, according to the process of formation. Another example is the classification of mass wasting phenomena according to Sharpe's designation as slow flowage, rapid flowage, landslides, and subsidence.[26] Such a distinction is usually very fine and is not always obvious from aerial photographs. But essentially the interpretation should take into account the processes, structures and stages of development of the landforms - an approach advocated by Lueder.[27]

146

It is therefore clear that the interpreter has to exercise his good judgement based on a sound knowledge of geomorphology, geology, climatology and other related subjects. Inevitably, features for one type of landform may be more easily interpreted than others; and between more contrasty environments, even the same image characteristic may differ in significance. Thus, Davis and Neal stressed the differences between humid and arid regions.[28] In the former, cohesive soils are usually associated with darker tones because of their higher moisture content, whilst in the latter, the clay surface of a dry playa often has greater reflectance than sand surfaces, and clay and salt have almost the same reflectance when dry. This fact necessitates some modifications of the photo-interpretation technique in analysing features of arid regions. Similarly, Verstappen has observed that 'photo-interpretation becomes less revealing in areas where only rocks of similar resistance outcrop or where the effects of selective erosion are obliterated'.[29] He stressed the importance of a knowledge of climatic factors such as rainfall, temperature, humidity, etc, and drainage pattern as essential for the correct evaluation of the lithology and structure of the rock on which the land-forms are developed. The climatic factors, most notably rainfall, affect the intensity of dissection of the terrain and the kinds of weathering process (chemical or mechanical) prevalent in the area, as evidenced by a contrast in the characteristics of the landforms developed in the temperate humid, tropical humid, sub-tropical and sub-arctic arid regions. Obviously, the climatic factors cannot be obtained from aerial photographs alone and have to be supplemented elsewhere. On the other hand, the drainage patterns developed over an area are more readily apparent, from which one may infer the nature of the rock and the kind of geological control at work. As early as 1932, Zernitz classified drainage patterns into six basic types and twenty four modifications.[30] A logical grouping of the more significant patterns, having taken into account their places of occurrence, is pictorially presented in Fig 4.2, which displays the characteristics of the drainage pattern developed over lowland areas (anastomotic, Yazoo, dichotomic, braided and recticular) and highland areas; and in the latter whether the drainage pattern has been

147

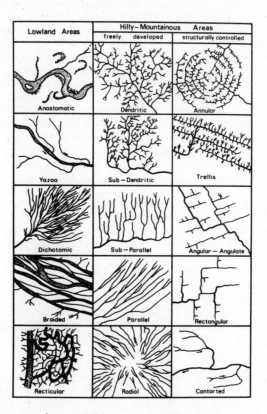

Fig 4.2 *Major types of drainage patterns (after Verstappen, 1963)*

developed freely (dendritic, sub-dendritic, sub-parallel, parallel, and radial) or with structural controls (annular, trellis, angulate, rectangular, and contorted). Of course, in actual situations, the drainage pattern developed over an area will not be so well defined and is likely to be a result of the mixture of two or more patterns. But this in conjunction with the grey tone criteria mentioned before will provide a strong basis for correct identification and evaluation of the images.

On the other hand, more detailed analysis of individual elements of the drainage pattern is possible using large-scale aerial photographs and involving the interpretation of river valleys and channels. It is possible to establish positively the type and size of the valley and its main elements such as slopes, terraces, flood plains, etc, and the less directly recognisable characteristics such as the marshiness of the slopes and the valley floor, soils, etc. All these add up to give some understanding of the work of the water. Kudritskii *et al* have given very detailed instructions on the methods of interpreting the river valleys under different terrain conditions, such as in mountainous, hilly and flat areas; and in different cover types and seasons such as with a forest cover and in winter.[31] This can lead to the identification of the various types of valley such as rifts, gorges, canyons; v-shaped valleys; trough-shaped valleys; box-shaped valleys; trapezium-shaped valleys; unpronounced valleys; dry valleys and ravines. Similarly, the characteristics of the river channels such as meandering, forking, predominant types of channel format-ions, etc, can be interpreted from aerial photographs. It should be noted that only some of these features can be interpreted from aerial photographs by direct identifi-cation clues and that both the qualitative and quan-titative aspects of the stereomodel have been exploited for the extraction of useful information.

So far, only the landform aspect has been stressed, but, as has already been shown, the identification of a land-form type leads to inference on the rock types and structural geology of the area, or, in some cases, even on the soil, vegetation and land use. Each rock type has its geomorphological expressions, which means that a strong correlation exists between the two. Some general observations can be made as a guide to photo-interpre-tation. Thus, landforms carved in igneous rocks exhibit the dominance of weathering along fractures, character-ised by a coarse and regular dissection with a strong tendency for the development of a dendritic drainage pattern owing to the absence of any structural influences. Resistance to erosion differs according to the chemical composition of the rock types, as the *basic* rock has a high concentration of silicates, resulting in the basic rocks being less resistant then the acid ones. On the other hand, landforms carved in metamorphic rocks stand out more prominently in relief as a result of the greater resistance of the rocks after metamorphism which also makes them less susceptible to differential erosion. Finally, for landforms carved in sedimentary rocks, very distinctive differences from those carved in the other two types of rock can be detected. The bedded rocks can facilitate differential erosion if they are of varying resistance; and their angles of dip strongly influence such landforms as escarpments or hog-backs; and if faulting and folding have occurred, structural valleys and ridges coincide with the anticlines and synclines respectively. In general, shales are weak and form depressions occupied by rivers whilst sandstones, conglomerates, etc are more resistant, forming ridges. Landforms developed on the same rock type will also differ according to the climatic environment in which they occur. In Plate 19 (Puckett Field area, Texas), an example is chosen to illustrate the impact of climate and rock type on landform. It is noteworthy that the horizontal rock strata combined with differential weathering and variations in tone caused by the fact that different rock types support different vegetation permit the identification of individual layers of rock. The dry river beds and the scrubby vegetation indicate an arid climate in the area.

As is to be expected, the success of photo-interpretation varies from one type of rock to another. Thus igneous rocks, especially the extrusive rocks, can be more easily interpreted, to be followed by sedimentary rocks, and finally, the most difficult of all, metamorphic rocks. A more detailed treatment of each of these can be found in a number of works, including Colwell, Verstappen, Lueder, Ray and Allum.[32]

2 *Land-use Analysis*. Land use is the use made of the land by man, and a knowledge of this is essential to an understanding and evaluation of man's impact on the natural environment. The bird's-eye view afforded by the aerial photograph is a great advantage for land-use analysis because it presents a holistic view of the spatial relationships of the different land uses within the limitations of the photo-scale.

There are two different levels at which land-use data

can be extracted from aerial photographs as a result of
the control by scales: (a) the more generalised level of
interpreting broad groups of land use from medium to
small-scale aerial photographs (1:10,000 or smaller)
which results in a thematic land-use map; and (b) the
highly detailed level of systematically determining the
use of each parcel of land in the area under consideration
from large-scale aerial photographs (eg 1:1,000) which
results in an inventory.[33]

Each of these two types of approach meets different requirements and faces different problems during interpretation. Obviously, for the more detailed level of interpretation, greater accuracy is required, and as a result
the interpreter must be well versed in the technique and
should possess a highly specialised knowledge of the
subject. There is also the need to stress the importance
of the cultural background because land-use interpretation from one cultural situation cannot be directly
transferred to another cultural situation without modifications. This is the same problem faced by geographers
in transferring Western models of developed countries to
non-Western developing countries; and photo-interpretation, to be accurate, must be based on an *a priori*
model developed for the area in question.

For this detailed inventory approach, the accuracy of
the land-use data interpreted depends heavily on the
classification scheme used; and the success of the classification scheme in turn depends on the clarity of definition and description of each land-use category. It has
been recognised that a single land-use classification
system which will serve all users at all times is impossible to establish. A classification scheme designed for
land-use survey on the ground is very different in terms
of detail from that for inventorying with large-scale
aerial photographs, since in the latter the more 'internal'
aspects of the land use will not be so readily visible. In
addition, there is a weakness in the land-use classification scheme caused by a confusion in the various concepts
about land resulting in mixing what is to be measured
and recorded. Land use has been variously viewed as the
human activity on the land, the natural qualities of the
land, land tenure, etc.[34] For geographical applications,
the land-use classification can be morphological or

151

functional or a mixture of both in design. The morphological scheme deals with the land cover rather than land use, as evidenced by the use of such terms as arable land, grassland, moorland, built-up area, etc - a good example being Stamp's scheme for the First Land Utilisation Survey of Britain in 1930,[35] and the more detailed two-level scheme for the Second Land Utilisation Survey under Coleman in 1960.[36] The functional scheme is activity-oriented, as evidenced by the use of such terms as agricultural, grazing, forestry, urban activities, etc. Anderson suggested that the latter may be more suitable as a general-purpose classification scheme for use with high-altitude photography or space imagery.[37] However, for a more detailed survey of land use, it has been recommended by Clawson that a single or pure line concept should be used in a single classification scheme and an inductive approach be adopted.[38] This means that the interpreter should interpret land use in as much detail as possible for the smallest recognisable parcels of land so that the uses can be grouped into the categories most appropriate to his own investigation (ie to maintain a high degree of flexibility).

As for the other more generalised level of land-use interpretation involving small-scale photographs, the problem of the classification scheme is equally applicable, although the demand for accuracy is less stringent. The number of categories of land use is obviously much fewer than in the detailed analysis case. The major source of error is due to grouping, resulting in some loss of detail caused by data aggregation within enumeration cells. A sampling design may also be employed to extract qualitative (as well as quantitative) land-use data along lines, at points or in areas on a random or stratified basis in order to produce small-scale land-use maps for large areas.[39] Berry and Baker stressed the suitability of the stratified systematic unaligned sample design for extracting land-use data at points from aerial photographs, and a general small-scale land-use map can be constructed based on the data at these points.[40] If this method is employed, sampling errors will occur.

These problems of extracting qualitative land-use data from aerial photographs at the detailed and generalised

152

levels will be better appreciated if we examine some actual examples of applications. As it is easier to interpret crop types than it is to identify types of retail establishments, a division into rural land use and urban land use is justified in the following discussion.

(a) *Rural Land Use*. Rural land use deals with such cover types as crops and vegetation in the natural environment and hence is concerned with such activities as agriculture and forestry. These cover types and activities are all directly recognisable from aerial photographs; and the photo-interpretation invovled is relatively straightforward.

(i) *Crop Types and Farming Types*. The identification of crop types in an agricultural area at different times of the year has special economic significance in a rural land-use survey. This can be carried out with large-scale (ie not smaller than 1:10,000) aerial photographs. It has been pointed out that the farm crops can be differentiated on aerial photographs by the unique *tonal* and *textural* qualities of their photographic images and by objects which are commonly found *in association* with them.[41] However, these tonal and textural character-istics of the crops tend to change at different growth periods, so that for the best results in the interpre-tation, aerial photographs should be taken three weeks before harvest.[42] Although tonal values registered by each crop on aerial photographs are unique at certain intervals of growth, they do not only reflect the con-ditions inherent in the growth of crops but also combine the effects of ground moisture, sky conditions and other photographic characteristics. The use of a *densitometer* has been particularly favoured as a means of screening out the undesirable effects of standardising tone measurement. Texture qualities also tend to vary with the external characteristics of the crops and the manner in which they have been planted, cultivated and harvested. Textures can be fine or coarse, but a more detailed breakdown into 'lined', 'plaid-like', 'corduroy', 'striped', 'swath' and 'mottled' textures was proposed by Goodman in the American context.[43] In Hong Kong, the differentiation between the paddy and the vegetables can be clearly made on textural characteristics alone

153

(Plate 16). The use of association of objects in
identification is an indirect recognising characteristic
which makes use of such relationships as the presence of
straw stacks in the field indicating grain crops and
the presence of hay stacks indicating hay. Similarly,
the presence of cattle or lanes leading to barns etc helps
to establish the identity of rotation and permanent
pastures. The cutting and tracking pattern in harvesting
can also help to distinguish many crops or crop types.

After the crop types have been correctly identified, a
more advanced stage is to identify the different types of
farming activities. Goodman has again shown that three
sets of criteria interpreted from aerial photographs serve
as good indicators.[44] These are (a) farmstead features;
(b) crop associations; and (c) the uses that are made of
corn and hay, of which 'crop associations' is the most
reliable indicator. This conclusion was derived from
study of the Northern Illinois and South Michigan
environment in the USA where four types of farming -
dairy farming, hog farming, cash grain farming, and hog
raising and beef fattening farming - can be identified.
An important observation made by Goodman was that the
farmstead features such as barns, granaries and silos
may change as the economic functions change, so that
farm structure and function do not necessarily correlate,
but as dairy farming is the highest type of farming with
a demand for elaborate farmstead structures, it is most
susceptible to being degenerated into the lower types of
farming if the economic conditions change. On the other
hand, the lower types of farming cannot be up-graded into
the higher ones without changing their existing farmstead
structures. This subtle point illustrates very well how
the logical process works in evaluating the significance
of imaging characteristics.

(ii) *Vegetation Survey and Mapping.* Vegetation is the other
cover type in the rural environment which shows less of
the influence by man, although some types of vegetation
such as trees may be artificially planted. But vegetation
is a kind of economic resource which needs to be
inventoried and classified. From the viewpoint of the
ecologist, all the components of the natural environment,
the rocks and soils, plants and animals, are defined in
terms of the vegetation components of the few principal

154

habitats, and the identification of these habitats is required in ecological research. This gives rise to the need for vegetation maps which may be regarded as a more specialised type of land-use map.

Aerial photographs have been found to be suitable for use in vegetation survey and mapping, especially over less accessible mountainous areas where the vegetation has been little disturbed by man, as demonstrated by the work of the Nature Conservancy staff in North Wales.[45] Various scales of photograph can be used which range from 1:5,000 to 1:70,000 depending on the output level desired. It is obvious that with the use of small-scale photographs only a very broad grouping of vegetation is possible. However, it is generally agreed that the optimum scale to use for the discernment of composition and boundaries of the vegetation is 1:10,000. Goodier and Grimes in their study of the Rhinogau in North Wales also employed 1:5,000 scale photography in order to resolve certain photo-interpretation problems that arose from use of 1:10,000 scale photography. The larger scale can allow more detailed information to be extracted, but reveals too much intra-community complexity. This preference for the smaller scale necessarily causes some lack of precision, but gives rise to a suitable degree of generalisation in mapping.

The accuracy of interpretation of the vegetation depends heavily on the skill and experience of the photo-interpreter, especially in recognising the different species. In general, the main identifying signs are texture, tone, shadow, and to some extent image form.[46] The importance of each varies according to the type of vegetation under interpretation. Thus, the textural pattern created by the image of tree crowns makes it easy to recognise forests and shrubs on photographs whilst the shadows of the forest and individual trees can give useful clues to identification. In the interpretation of different types of grass, tone plays an important part, whilst tone and texture are equally useful in dealing with marshes. Other clues such as heights and sizes may also be employed together to help identification. All these characteristics can be summed up in table form for use in identifying a specific type of vegetation at a specific photo-scale. Kudritskii *et al* have illustrated

an example for identifying trees - spruce, fir, pine,
birch and aspen - using nine identifying signs: (a)
overall image tone of the stand, (b) shape of crown, (c)
tone difference between illuminated and shaded parts, (d)
height variations in crowns on the main canopy, (e) crown
distribution, (f) variations in crown diameters of the
main canopy, (g) vertical extent of crowns, (h)visibility
through canopy, and (i) characteristics of a single tree
shadow discernible in sparse stands, from aerial photographs at two scales: 1:15,000 and 1:25,000.[47] This may
be regarded as a general photo-interpretation key since
it has no reference to a specific location. This can
also be compared with a similar key for herbaceous plant
communities of the Rhinog Mountains produced by Goodier
and Grimes on 1:10,000 scale panchromatic air photographs.[48] But, as the authors admit, general ecological
experience has played an important part in the interpretation. However, it appears to be quite useful to follow
a key in the interpretation if it is properly designed.
Sayn-Wittgenstein has employed crown characteristics only
for the recognition of tree species on large-scale aerial
photographs and has produced as a result an elimination
key.[49] By crown characteristics, such features as crown
shape and branching habit are involved. Very large scale
aerial photographs such as 1:600 were used in order to
see these clearly. The crown shape can easily be
described as oval, conical and cylindrical. But other
factors such as shadows, tone and a knowledge of the
ecological and silvicultural characteristics of species
are also employed for the recognition of tree species.
Thus, some of the identifying characteristics may be
rather indirect.

The vegetation mapping carried out in different highland areas of Britain by the Nature Conservancy staff
has shown that, with the use of 1:10,000 photographs,
the dwarf shrub communities, the mire communities and all
the bog, moor and heath vegetations can be easily
identified whilst grass and tall herb communities on the
whole are intrinsically difficult to interpret as a
result of different management practices which can
markedly affect their appearance. This points to the
fact that with more impact by man the identification of
vegetation types may be difficult, probably because the
ecological system has been disturbed, resulting in a

mosaic of very complicated plant communities, as is evident in the lowland areas of Britain.

In their study, Goodier and Grimes also compared the different approaches to vegetation mapping using aerial photographs. This really is a question of how to derive the classification of units to be mapped. There are the *a priori* classification which involves working out a classification in the field beforehand and then making one's photo-interpretation comply with this; the *a posteriori* classification which requires one to delineate units on the photographs before going into the field to sample the marked units and classify them; and finally a specially designed classification suited for photo-interpretation purposes. It appears that the third approach is the most practicable. It has also been realised that the mapping units cannot be defined solely on the basis of species presence and physiognomy, but should also reflect the visual effects of geomorphology, soil type and moisture content on the vegetation, thus favouring the so-called landscape method of interpretation in Russia.[50] An example of the vegetation map of a part of Endrick Water in Loch Lomond National Nature Reserve, Scotland, is shown in Fig 4.3 which should be compared with the corresponding photographs (Plate 20).

(iii) *Changes in Rural Land Use.* Land use is not a static phenomenon and is undergoing changes all the time to reflect the different degrees of the impact of man on the environment. Sequential aerial photographs have found many applications in this field. Dill has demonstrated a straightforward case of comparing agricultural land-use changes between two time periods, 1944 and 1955, by employing aerial photographs.[51] Two different types of photograph were used: contact prints at the scale of 1:20,000 and the air photo index sheet which was in fact a photo-mosaic covering a much larger area. It was possible to obtain more detailed information from the 1:20,000 contact prints for a restricted area. The usual procedure was to identify those areas of change and to mark them on the later set of aerial photographs. These were then annotated and their areas measured using a transparent grid. In the case where air photo index sheets were used, they had to be studied under magnification in order to identify and outline

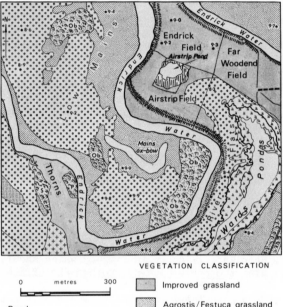

VEGETATION CLASSIFICATION

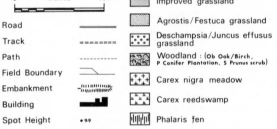

0 metres 300	Improved grassland
	Agrostis/Festuca grassland
Road	Deschampsia/Juncus effusus grassland
Track	Woodland : (Ob Oak/Birch, P Conifer Plantation, S Prunus scrub)
Path	Carex nigra meadow
Field Boundary	Carex reedswamp
Embankment	Phalaris fen
Building	
Spot Height •9.9	

Fig 4.3 *Vegetation map of a part of Endrick Water in Loch Lomond National Reserve, Scotland (Department of Geography, University of Glasgow)*

land-use changes on them. Avery similarly advocated such an approach to study land-use changes in Clarke County, Georgia, USA between 1944 and 1960 by means of 1:20,000 United States Department of Agriculture aerial photographs.[52] He pointed out that there were only six types of land use which could be *consistently* recognised from aerial photographs and therefore best suited for comparing changes over two or more distinct points of time: (a) cultivated land, (b) pine forest, (c) hardwood forest, (d) urban land, (e) idle land, and (f) water. The results clearly revealed that the county had already been effectively transformed from an agricultural to an urban-industrial region by 1960.

(b) *Urban Land Use.* Urban land use reflects man's use of an artificial environment - the city where there is a concentration of people and a focus of human activities. Most of these activities are under cover and hence are not directly visible from aerial photographs. Also, the diversity of these human activities causes greater difficulty in photo-interpretation than in the case of rural land use. But aerial photographs have been and still are favoured by urban geographers and town planners as a major source of urban information at varying levels of detail controlled by the photo-scale. Indeed, for more general uses, aerial photographs provide the necessary background data on the geographical setting of the urban environment. Thus, Witenstein incorporated this as a typical approach in the sequence of inventory, analysis and plan to be adopted by planners.[53] He illustrated the usefulness of such an approach with an example from the United States - the town of Rockville, Maryland, where a planning problem to expand the facilities of the existing Central Business District to meet an unexpectedly large population overspilled from the nearby metropolitan area was in hand. His sequence of procedures involved first of all the examination of small-scale photo-mosaics to obtain an overall view of the urban environment; then stereopairs at large scale (by which he meant photographs at 1:20,000 or larger) were used to study each part of the urban area intensively. Gross land-use types, such as residential, commercial, industrial, transport, vacant land, etc, were interpreted and mapped from the aerial photographs separately as a set of overlays on the photo-mosaics.

159

For each type, measurement was made of the size and capacity of use, thus obtaining statistical data (observed variables) on an areal basis. Computation could then be made of the ratios of land used, land zoned, service availability and facility accessibility to residential, commercial and industrial needs, etc, thus yielding constructed variables for analysis and planning, with the two other procedures to follow in the sequence. Witenstein has carried out a ground check which supported the high accuracy of the photo land-use data and the usefulness of aerial photographs in developing detailed statistical data.

The approach to mapping gross urban land use is similar to that used in rural land-use mapping with the need to construct beforehand a classification scheme defining the types of land use to be mapped. This classification scheme is in fact a photo-interpretation key or a typology to aid the identification of urban land-use types.[54] In residential, commercial, industrial and transport uses, the types and arrangement of structures are employed as clues to photo-identification.

So far, Witenstein's approach cannot provide detailed urban land-use data because the photo-scale (1:20,000) employed is not particularly large. In order to meet this need, large-scale photographs of 1:10,000 and above are really required, and the photo-interpretation has to be done with the aid of a key carefully prepared on the ground with reference to a representative sample area.

A modern piece of work by Collins and El-Beik well illustrates this method of extracting urban land-use data from large-scale aerial photographs.[55] A set of mediocre-quality 'split vertical' aerial photographs covering the city of Leeds at the scale of 1:10,000 was used. The first step was to prepare a photo-interpretation key. A small sample of photographs covering areas with major urban land-use types was selected and a land-use survey was carried out in the field. The land-use data so obtained were mapped on trace overlays of these photographs. From this, the urban land-use key could be compiled. The key which resulted embraced eight broad types: commercial, industrial, residential, bodies of water, transport, public buildings, open improved land

160

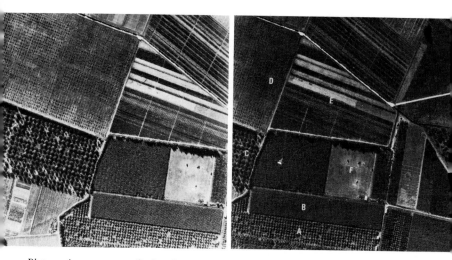

Plate 17 A stereogram of a farming area in Kent, England. Texture has played an important part (in addition to height) in aiding the identification of different types of crops and farming activities: (A) mature small fruit trees, eg plums, (B) hopfield, (C) mature large fruit trees, eg apples, (D) young fruit trees, (E) vegetables, and (F) poultry range on a previous hopfield *(Hunting Surveys Ltd, Boreham Wood)*

Plate 18 A stereogram of a steel plant in Scunthorpe, England (taken with a Wild RC8 camera with a focal length of 152.18mm at an altitude of 488m). Height and size are useful clues to the identification of the various circular objects: cooling towers (A), gas storage tanks of varying sizes (B, C), and concentration tank (D). Note also the coal dumps (E), conveyor belts (F) and the heavily polluted atmosphere *(Hunting Surveys Ltd, Boreham Wood)*

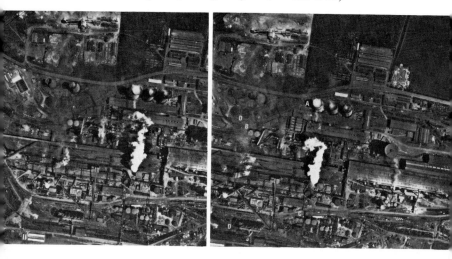

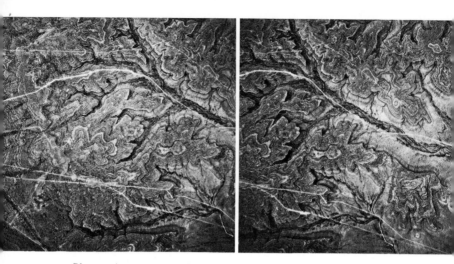

Plate 19 A stereogram of the Puckett Field area in Texas, taken with a Wild RC9 camera with a focal length of 88.5mm at a flying height of 2,125m above the ground. Note the horizontal rock strata clearly marked by alternating layers of vegetation. The scanty and scrubby vegetation is particularly indicative of a dry environment. It is possible to construct an accurate geological map of the area based on this stereogram *(Wild Heerbrugg Ltd and Allied Engineering Co, Topeka, Kansas)*

Plate 20 A stereogram of Endrick Water in Loch Lomond National Nature Reserve, Scotland, showing a meander and an oxbow lake near its estuary. The large-scale aerial photographs (about 1:1,000) allow detail interpretation of vegetation types in a low-lying marshy area. Very low sun angle is obvious at the time of photography, which gives long shadows (thus helping the identification of vegetation species) and emphasises field boundaries and furrows. Textural changes within different vegetation types are extremely distinct, thus facilitating the mapping of their boundaries. The resultant interpretation based on this stereogram is shown in Fig 4.3 *(BKS Surveys Ltd)*

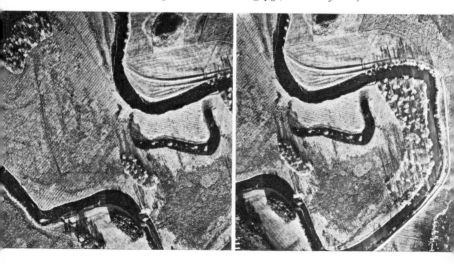

and open unimproved land. For all types except open unimproved land, further distinction into more specific uses was made. Thus, the commercial use was subdivided into offices and shops; the residential use into detached houses, semi-detached houses, terraced houses, back-to-back houses, single-storey houses and blocks of flats; the transport into railways, roadways and waterways; and the public buildings into educational buildings, hospitals and churches. In addition, an industrial land-use key was prepared to give a detailed classification of industrial types. A detailed description of each item with accompanying terrestrial and aerial photographs and stereograms was given in the keys.

Once the interpreter had shown consistent and accurate identification of the various land-use types in the sampled aerial photographs, the key was extended to the rest of the city. Thus, all the land-use units in the key that could be identified were traced on an overlay. This was then directly transferred to the base map which was roughly at the same scale as the aerial photographs. Any unidentified units from 'problem areas' were ground-checked and a 'special problem area' key was then compiled for future work. In this way, the level of reference of the interpreter was gradually being raised to yield better results.

Collins and El-Beik have shown that it was possible with this approach to identify correctly 88.5 per cent of the total units of urban land use with an underestimation of 5.2 per cent. They pointed out in particular the difficulties in identifying residential buildings. These were mainly caused by the fact that in the city of Leeds many buildings had changed their function through time but not their form. But better-quality and larger-scale photographs would have consistently given more accurate results. They concluded with a suggestion that a photo-scale of 1:2,500 would be optimal for urban land-use studies.

Aerial photography is not only confined to general land-use surveys but is also frequently employed in more specialised urban studies such as (i) transport, (ii) townscapes and (iii) changes.

i *Transport*. The flows of vehicles and pedestrians along the artificially created urban maze of streets and roads represent the dynamic component of the urban environment, and have given rise to the most conspicuous, immediate and recurrent problems in the urban system, which are the major concern of urban geographers and planners. The Highway Research Board in the United States has made the most concerted effort in tackling these transport problems by means of different approaches, in which the use of aerial photographs has assumed an important role. These applications, in most cases, entail the use of large-scale photographs from which the following information on the traffic flow pattern can be extracted: (a) vehicle types, (b) density, speed and direction of flows, and (c) location and availability of parking facilities. Therefore, the operations involved are essentially vehicle identification and vehicle counts, both of which require a combined use of photo-interpretation and photogrammetry. The main advantage of aerial photography, as has already been pointed out, lies in its ability to freeze urban movements at a specific time, thus providing an overall picture for detailed analysis.[56] In fact, traffic data are always used in conjunction with urban land-use data in transport planning which is an important phase of urban land-use planning.

ii *Townscapes*. The urban environment impresses upon man with its own townscape just as the rural environment expresses itself with its landscape. This is the result of the arrangement of buildings (the living 'shells' of man) of different types and styles. Objectively, aerial photographs, usually at large scales, record very well the qualitative characteristics pertaining to the buildings, such as roof forms, façades, etc. The different types and styles of building are then identified and viewed in relation to one another 'so that a comprehensive assessment of planning development, and stylistic criteria can be made rapidly and simultaneously'.[57] Thus, these qualitative data are most useful in understanding the course of townscape evolution from an architectural point of view, as already explained when the metric data from aerial photographs for this kind of urban morphological study were considered. The different stages of growth of a town can then be traced.

Today, 'townscape' may be studied as a subjective element reflecting an inhabitant's perception of his living
environment, as exemplified by Lynch's 'mental maps'
representing one's image of the city.[58] The use of
aerial photographs in this connection is more strictly
descriptive and involves psychological considerations.
Dale's interesting investigation of children's reactions
to maps and aerial photographs has revealed the importance
of past experience in forming a picture of the area in
the minds of the children.[59] On the other hand, non-
conventional aerial photography may be applied to present
non-Euclidean relationships of certain aspects of the
urban environment, such as the panoramic photograph whose
panoramic distortion results in a picture which appears
to depict the distance decay function from the city centre
towards the periphery. The panoramic photograph of
Downtown Boston (Plate 21) resembles Hägerstrand's
logarithmically transformed map of Sweden centred on
Ashby depicting distance decay. (Fig 4.4)

(iii) *Urban Change*. The urban environment is subjected
to more rapid change than its rural counterpart. This is
not surprising, as the urban environment is a microcosm
of human activities. Such changes may be physical (eg
townscape), social (eg land use) and economic (eg
transport), and are significant indications of the
development trends of the urban environment. If aerial
photography is available for a city over regular periods
of time, a series of static land-use patterns and dynamic
movement patterns can be generated through descriptive
photo-interpretation. From these maps, quantitative
measurement of separate land-use types can be made and
any persistent trends can be detected. From the town-
planning point of view, sequential photography supplies
information for projecting the future growth of a city,
and the appropriate course of action necessary to
maintain orderly urban growth can be formulated. Thus,
Howlett employed aerial photographs to detect urban
growth and changes due to the impact of the communication
systems in the United States.[60] Similarly, Wagner
measured on aerial photographs changes in land use around
highway interchanges where access to both free and toll
roads was easy.[61] These land-use changes generated a
great amount of traffic which might cause an interchange
to be congested and prematurely obsolete. Aerial photo-

163

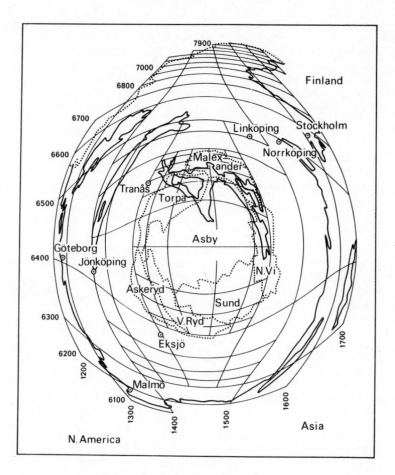

Fig 4.4 *A map of Sweden in an azimuthal logarithmic projection*

164

graphs taken at approximately one-year intervals covering
a period of five years were used, and altogether 31 ton-
toll interchanges and 33 toll interchanges were studied
in eight states from which generations could be formulated. An arbitrarily fixed area within a 3-mile (4.83km)
diameter circle centred on the point of intersection of
freeway with the cross road was studied; and the land use
that occurred within each one-mile (1.61km) zone was
interpreted from aerial photographs. The changes were
identified, outlined and labelled after a comparison of
corresponding photographs at different time periods. The
area for each type of land use was measured by the dot-
grid method. Field checks revealed rather accurate
results of interpretation.

Similar approaches to studying land-use changes in a
city were reported. Falkner made use of time-lapse
aerial photography in determining short-range (1960-66)
land-use changes of a small district,[62] whilst Richter
strongly advocated, from the economic point of view, the
use of sequential aerial photography for planning the
long-term growth of small urban communities with 50,000
population and below, illustrating his point by a series
of land-use maps that he produced from aerial photographs
for the city of Janesville, Rock County, Wisconsin for
the years 1940, 1950, 1956, and 1963.[63]

By Indirect Observation

The applications in this group have made use of photo-
interpretation techniques to yield data relating to those
phenomena of the terrestrial environment which are by
themselves not directly apparent on the photographs.
These phenomena can be natural or artificial. The technique requires the application of one's power of deduction
and generally involves establishing a correlation between
the obscure phenomenon and an obvious feature on the
photograph. Therefore, the presence of that feature
serves as an indicator for the obscure phenomenon in the
area. This really makes use of the 'proxy' or 'surrogate'
concept of soil science where soils are mapped on the
basis of the association between soil series and natural
vegetation communities.[64] Obviously, the correlation can
only be established by an empirical approach requiring
extensive field experience in the subject. Thus, this

method is also known as *correlative photo-interpretation*. There are two major fields of application corresponding respectively to the natural and artificial environment which are worthy of some detailed discussion: (1) soil mapping and (2) urban analysis.

1 *Soil Mapping*. Soil is that part of the regolith in which plants grow. Dokuchaev, the Russian 'Father of Soil Science', viewed it as an independent body with a natural history living its own special life which is intimately related to climate, vegetation, parent rock, relief of the region and its age.[65] The interaction of these soil-forming factors in different combinations results in different soil types possessing different properties. It should be pointed out that the development of soil is not only horizontal but also vertically downwards so that an individual soil is really 'a three-dimensional piece of landscape'.[66] This downward development may extend to a depth of 1.3m to 1.8m or even more.

An understanding of the spatial distribution of the individual soil types over an area is a major objective of soil studies, especially because of its economic importance as far as agricultural development of a country is concerned. Therefore, a major task is to carry out soil mapping by which individual soil types are identified and their separating boundaries are drawn. Because of the spatially rapid changes of soil type, soil mapping is difficult, and for a detailed soil map at a scale of 1:10,000, the majority of mapping units resulting from ground survey are only soil complexes which mean that certain soils are more likely to occur within the unit boundary than are others.

The method of soil mapping is based on a study of the soils and the soil-forming factors, and this can be usefully applied in conjunction with aerial photographs for the purpose. It was discovered by the Dokuchaev Soil Institute in the USSR that data obtained by aerial photography were of great importance both for 'detailed' and for 'intermediate'-scale surveying. The 'detailed' soil maps are usually at the scales of 1:10,000, 1:25,000 and 1:50,000, whilst the 'intermediate'-scale maps are usually compiled at a scale of 1:100,000 or 1:200,000.[67] By topographic mapping standards, these are really medium to small-scale maps; and the advantage of small-scale aerial photography obviously lies in producing a spatial generalisation on the complex series of soils over an area. As in the case of vegetation mapping mentioned in an earlier section, the 1:10,000 scale photography is the uppermost scale in the range to be used for soil mapping and is considered to be the best to use, since it will avoid the excessive detail of the larger-scale photography without losing accuracy. This special requirement of soil mapping for a suitable degree of generalisation should be noted and indeed affects its depth of soil interpretation from the photographs.

When aerial photographs are to be used, one must realise the fact that the bare soil surface is not always visible and is often covered by plants. The necessity to take account of downward development is another difficulty. Therefore, the interpretation of soils is normally an indirect exercise requiring inferences to be drawn from other visible features. Fortunately, the most important soil-forming factors are plant cover and relief form, which are clearly and accurately recorded on aerial photographs. It has been pointed out that even aerial photographs at a scale of 1:25,000 can give an image of different plant associations and of microrelief.[68] These are the two indirect identifying signs for soil interpretation. Thus, for the relief forms, Kudritskii *et al* gave the following general examples of associations:[69]
(i) barchans and dunes - dry and sandy soils;
(ii) vertical walls, deep, narrow ravines and gullies with jagged edges - solid sandstone and loess; and
(iii) precipices and sliding slopes with salient formations and smoothed outlines - clayey soils.
For the vegetation cover, one finds that:
(i) pine forests are typical of sandy soils;
(ii) spruce and fir predominate on clayey and loam soils, with spruce usually growing on low, marshy terrain;
(iii) osier usually grows on sandy soil and silt; and
(iv) meadow vegetation occurs on alluvial sand, sand-silt or peat-silt soils.

These two basic indirect identifying factors can be combined to give a multiple correlation of the relation-ships for more positive soil identification. Thus, in interpreting the soils of the Caspian lowland as reported by Simakova, the occurrence of a special relief form called 'liman' which is a large topographic depression with a uniform soil cover and which contains meadow vegetation, is indicative of the presence of meadow soils. More specifically, one can distinguish three types of liman accordง to Gerasimov, namely, (a) open limans which are in the main path of flow of river water and melted snow; (b) half-closed limans which are connected in times of high river-floods to the main bed of the liman or having an outlet for snow melt waters; and (c) closed limans or snow limans which are not connected to the fluvial network.[70] The different moisture conditions in each type determine vegetation and soil type. Even for open limans several levels can be identified depending on the amount of flood water each level receives. The main level which is flooded nearly every year is covered with couch-grass meadows (*Agropyrum sp*) with the formation of heavy loamy soils such as meadow solonetsous, meadow-carbonate, and meadow solonchakous soils. In the sunken areas of the main level where water stagnates, the bogged-up meadow-solodised soils characterised by a certain degree of glei formation are found, which are associated with the plant cover of sedge. At the higher level where flooding is less frequent, a type of grass called *Atropis distans* occurs, indicating the presence of solonchak-meadow soils.

The other two classes of liman are therefore very similar to these different levels of the open limans. Thus, the half-closed and closed limans generally develop solodised meadow soils as in the main level of the open limans with the predominance of couch-grass as the plant cover.

This special example of soil mapping in the lowlying Caspian area is particularly complicated but illustrates very well the type of data that can be extracted from aerial photographs to help the identification of soil types. Altogether fourteen types of soils and soil complexes have been established, based mainly on the photo-interpreted characteristics of vegetation and relief form where such soils occur. The density of the vegetation (or percentage of the ground covered) formed in each type of soil has been found to be a useful indicator too.

In the case where the soil surface is directly visible
on the photograph, the major identifying sign will be
the *tone* or the spectral reflectance of the different
surfaces which can be directly measured by means of the
microdensitometer to produce the characteristic spectral
reflectance curve for each type of surface (Fig 4.5),

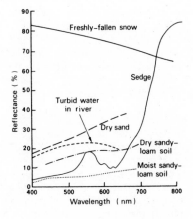

Fig 4.5 *Light reflectance curves for
different types of terrestrial surface
(after Simakova, 1964)*

as already discussed in the section on quantitative
photography in Chapter III. On the other hand, even if
the soil is covered by vegetation, the vegetation cover
itself provides a surface for similar spectrophotometric
measurement; and although the tone is directly related
only to the type of vegetation, nevertheless it indirectly helps to identify the underlying layer of soil.

Another useful approach in soil mapping is to make use

of the soil *patterns*, which result from the repetition of
a particular disposition of forms and colours and are
particularly visible to human eyes. Such an approach is
most effective in an area which is not covered by crops.
Evans has been able to demonstrate the usefulness of
such a method for soil survey in lowland England and has
been able to identify (i) the coastal and fen patterns
and (ii) the upland patterns.[71] In (i), it has been
possible to identify the former estuarine marsh creek
system characterised by a dendritic drainage pattern,
sinuous channels with calcareous, micaceous, very fine
sandy loam to silt loam, and the former tidal marsh
creek system characterised by complex interlocking
dendritic drainage patterns with lighter-toned calcareous,
micaceous silt loams. In (ii), the patterns are
associated with a periglacial or glacial environment
including stripes and polygons, circular and vermiform,
developed on soils in glacial drifts. All these soil
patterns enable the photo-interpreter to assess soil
variation quickly within the mapping unit and to under-
stand the causes of variability.

There have been some attempts to test the reliability
of the soil boundaries drawn as a result of photo-
interpretation by the direct or indirect methods. One of
these is a numerical method described by Webster and
Wong, using the Upper Thames Valley to the west of Oxford
as an example.[72] It must be noted that this test area
consists of a lowlying plain with low rounded hills or
plateaus and widely spaced drainage lines. The maximum
relief is about 30m, and the area has no appreciable
extent of natural vegetation. The photo-interpretation
of soils made use of panchromatic black-and-white aerial
photographs at a scale of 1:20,000. The results were
checked by regular sampling along a linear transect
(6km long) which was chosen randomly across part of the
area. The soil was examined at regular 20m intervals
along the transect and the soil properties at each of
these points were assessed quantitatively in the field
from samples taken with a 10cm auger to a depth of 1m.
In all, 294 points were sampled in this way. Ideally,
a soil boundary should separate two distinct types of
soil so that one tends to recognise boundaries that
correspond to maximum rates of change of soil properties
with respect to distance. These boundaries are really

lines of contrast or inflexion in mathematical usage.
The testing of the accuracy of soil mapping from aerial
photographs therefore involved comparing the inflexion
points discovered along the line of transect on the
ground with those interpreted along the same line from
the photograph. Since the field sampled results dealt
with a large number of properties of the soils at each
point, a multivariate method called principal component
analysis was used to group these into components of
characteristics; and the first principal component which
accounted for 31 per cent of the total variation was used
as standard for comparison. A visual comparison could be
carried out, but for more objectivity, a statistical
Student's t test could be used, based on the argument
that the best position for the boundary would be the
one that maximised the differences between and minimised
the variances within classes of sampling points on either
side of each line. The comparison both by the visual and
statistical methods indicated that the agreement between
the ground and photo-interpreted results was very good
indeed, although some omissions of boundaries were also
noted. The objective statistical comparison gave the
calculated inflexions even closer to the air-photo-
interpreted boundaries than their visual estimates. When
the photographs were re-examined, it was discovered that
these boundaries were omitted mainly because they
occurred without any perceptible associated change in
landform. This indicates the major weakness of the
indirect method of photo-interpretation, ie that the
correlation (in this case between relief form and soil)
is not 100 per cent accurate. Also, in an area of flat
terrain, the interpretation is more difficult as the
relief form is not distinct as compared with an area of
higher relief, and consequently, more field experience
is required for more accurate photo-interpretation.

2 *Urban Analysis.* The problem of applying aerial
photographs to urban analysis is much more complex than
in the case of soil mapping although the underlying logic
is the same. This is because the major concern of urban
analysis is the extraction of socio-economic data which
are not directly observable on aerial photographs. These
have to be related to the physical elements, such as the
'proxies' in soil mapping. As early as 1948, Branch
had already pointed out that 'although photographic

representation is of the physical corpus and does not reveal economic, sociological, or governmental material directly, a surprising amount of indirect information pertaining to these fields is reflected in the three-dimensional characteristics of the community'.[73] Green, a sociologist, soon provided the rationale for the existence of such a relationship.[74] He argued that since the city comprises both a physical system having physical structure and a social system having social structure, the two components are not logically separable. It follows that human groups occupy the physical space and facilities as a result of their interactions and adjustment to the environment. Thus, the directly observable physical data on the aerial photographs should have meaningful sociological correlates. Such an argument is in fact based on Park's ecological theory, which can be traced back to Darwin's concepts of the intimate interrelationship between organism and organism as well as between organism and environment. One also sees the influence of Chombert de Lauwe's social space concept mentioned earlier.

The establishment of this socio-physical relationship is an essential step in this method of correlative photo-interpretation. Such a relationship can be established either empirically (inductive) or theoretically (deductive). The empirical method requires the use of ground survey to associate a socio-economic variable with the visible elements imaged on the photographs. The theoretical method, on the other hand, makes use of known functional relationships, which have been quantitatively established beforehand and can be tested for consistency and reliability by statistical methods. This indirect method of photo-interpretation is generally known as 'inventory-by-surrogate' in the United States and has been gathering momentum under intensive investigation and refinement by urban geographers who are interested in using remote-sensing techniques.

To exemplify this approach, the classical study by Green is best examined in greater detail. Green's work was aimed at studying the social structure of the city, which involved a study of residential sub-areas of several American cities. His approach started with the extraction of the following types of data from the aerial photo-

172

graphs: (1) the location of residential areas relative to three concentric zones with mid-point in the Central Business District (CBD), ie using Burgess's concentric zone model as a framework, and the identification of these three zones based on major breaks in land use and changes in building types; (2) the description of the residential areas in terms of the internal and adjacent land usage, from which an ordinal scale of 'residential desirability' (namely, favourable, neutral, unfavourable) was established; (3) the prevalence of single-family homes, ie a count of the single-unit, detached-type dwellings, thus giving another ordinal scale of high, medium, and low occurrence; and (4) the density of housing in average numbers of dwellings per block, giving rise to a third ordinal scale: high, medium and low density.

The second stage is to carry out ground checks to determine the accuracy of these photo-interpreted data, and the results showed that (a) up to 99 per cent of the residential structures were correctly identified; (b) the total number of individual dwelling-units in all types of structure was under-estimated by 7 per cent; (c) the overall average density of dwelling-units per block was under-estimated by 1.7 per cent; and (d) the detached, single-dwelling units were over-estimated by 5.3 per cent. The conclusion was that a comparatively accurate picture of the relative structural character-istics of the residential areas had been obtained.

In the third stage, the residential areas were ranked according to the two indices: (a) the prevalence of detached, single-dwelling units, and (b) the density of dwelling-units per block, first with the photo data and then with the ground data, and these two sets of ranking were compared by means of Spearman's Rank Correlation. It was found that near-perfect correlations of 0.98 and 0.99 respectively were obtained for these two indices used, thus proving that the photo data were just as valid as the ground data.

In the fourth stage, the sociological characteristics of these residential areas from other statistical sources were studied to see whether any of these were in fact related to the above categories of photo data or not.

173

By statistical method, it was possible to strongly establish the existence of such a socio-physical relationship. For example, it was confirmed that significant differences between zones in the city occurred in educational status, average monthly rentals and adult crime rates, supporting the validity of the classical ecological model. Another observation was the consistency in correlation between the prevalence of single-family houses and the socio-economic status of the urban sub-areas in terms of occupational status, education, income, rental values and ethnic composition. There was still another important finding in that correlation coefficients between average number of dwellings per block (by census tracts) and average number of persons per block were consistently of the order of 0.95 and above. Thus, it is possible to estimate the number of people living in the residential neighbourhoods from aerial photographs.

Finally, as the predictive value of a single category of photo data is limited, these social and physical data were combined by means of a multivariate technique - the Guttman Scale - by means of which the several qualitative variables were combined to form 'a single continuum defining the exact position or rank of each object (or sub-area) in relation to every other object in the sample'. Thus, a hypothetical continuum of a scale of 'residential desirability' of American cities was constructed, based on the four photo-data categories. A second scale, the 'socio-economic status scale', was also constructed using the five social data items: (a) median annual income, (b) prevalence of within-dwelling crowding, (c) prevalence of home ownership, (d) prevalence of social disorganisation, and (e) educational achievement. Thus, each of these scales provides an empirical definition of common characteristics among urban spatial patterns, with one referring to physical structure and the other to social structure, and these two scales were strongly correlated (with a product-moment correlation coefficient of 0.88 at 0.001 level of significance), indicating the strong power of the aerial photographic type of physical data in predicting the social structure of the city. In one case, the residential desirability scale was found to be able to account for 78 per cent of the variation in socio-

174

economic status distribution.

It is notable that Green's work was applied only to the
residential sector of the urban environment which is in
general more uniform in socio-economic characteristics
and hence shows a higher degree of correlation between
structure and function. Another point is Green's
reliance on Burgess's concentric zone model as the
initial framework for photo-interpretation, which may
be outmoded today as such an ecological model is static
and cannot take account of the temporal changes in the
patterning of the city. Such an approach will obviously
meet with great difficulties when applied to non-Western
cities, especially those in the developing countries
where a dual economic structure incorporating both the
traditional and the modern economic characteristics
has been observed.[75] The result of such a form of
development is highly mixed usage of the urban environment with the minimum correlation between physical
structures and functions. However, Green has demonstrated, to urban sociologists at least, in the fashion
of Chombert de Lauwe, the usefulness of aerial photographs as a source of qualitative urban data, although
he has not made clear what scales of aerial photographs
are most suitable for work of this kind.

The pioneering work of Green has initiated an important
line of applications of aerial photographs in urban
analysis, notably housing quality studies. Wellar has
found that in American cities a variety of features which
correlated well with the housing quality were identifiable on the aerial photographs. These included building
frontages, open or vacant land, unpaved versus paved
lots, amount of on-street parking, proportions of multi-
family and single-family dwelling in the area, architectural style, landscaping, conditions of lawns, presence
of litter, and lacking of curbing along parkways.[76]

The same technique has also been applied to extract
socio-economic data on the urban population. Mumbower
and Donoghue have demonstrated an approach with large-
scale aerial photography (1:10,000) to delineate urban
poverty by correlating it with such identifiable features
as structural deterioriation, debris, clutter, and lack
of vegetation, walks and paved streets.[77] They

discovered that urban poverty was closely associated with residential areas located adjacent to the CBD, industry and major transport arteries. These were also found to be strongly correlated with low income, unemployment, low education achievement, family crowding, crime, low health status and lack of community facilities.

More recently, Metivier and McCoy illustrated an approach with the use of large-scale black-and-white aerial photography (1:6,000) to map the urban poverty areas in Lexington, Kentucky.[78] They found that house density was the most significant criterion for the identification of substandard housing on the photographs. This was expressed as the number of houses per acre for each block, which could be averaged and mapped for each census district. It was also found that this house density could serve as an index of socio-economic conditions, and the results of the analysis revealed that both family income and average house value had a strong negative correlation with housing density.

CONCLUSIONS

In this chapter, we have examined the advantages of aerial photographs as descriptive models which are not always separable from their metric counterparts. The technique of photo-interpretation which has been basically viewed as a classificatory process is particularly geared towards the extraction of these qualitative data relating to our terrestrial environment; and it should be reiterated that these qualitative data are in no way inferior to the metric data acquired by means of photogrammetric techniques. It is only the undue concentration on one single aspect of aerial photographs, qualitative or metric alike, that a geographer should try to avoid if he wants to make best use of the stereomodel as a reduced-scale replica of real-world phenomena. On the other hand, the general method of photo-interpretation is applicable to both the physical and the human environment, although specific modifications have to be made depending on the scale of the photographs used and whether directly observable phenomena are dealt with. Indeed, a practical contribution towards applied geography has been made possible today by blending the physical and human characteristics together for an integrated approach such

176

as land evaluation, to be discussed in the final chapter.

V THE USES OF NON-PANCHROMATIC AND COLOUR PHOTOGRAPHY

The various applications of aerial photography in geography so far discussed have centred mainly on one standard monochrome film type - the black-and-white panchromatic film which is equally sensitised to the whole visible spectrum with wavelengths ranging from below 400nm to about 700nm. This makes it suitable for general-purpose uses. There are other monochrome films such as the orthochromatic film which is sensitised for wavelengths from below 400nm to about 600nm and the infra-red film which is sensitised for wavelengths from below 400nm to about 900nm, as discussed in Chapter II (Fig 2.8). All these may be regarded as special-purpose films because they are specially sensitised to some particular wavelengths so that they can record some objects more distinctly than others. This usually means that the photographic tone contrast or difference in brightness between the image and its background is enhanced, thus allowing the image to be more easily interpreted. Apart from these monochrome non-panchromatic films, there are also colour films which are capable of depicting our terrestrial environment either in true colour, which is more familiar to the human eye, or in false colour which attenuates the effect of colour on certain objects at the expense of others, to facilitate detection. All these different types of special film have significant impacts on the geographical applications of aerial photography today.

ELECTROMAGNETIC SPECTRUM: THE VISIBLE PORTION

The electromagnetic spectrum is an effect produced when electromagnetic radiations are resolved into their constituent wavelengths or frequencies. Within the spectrum the radiant energy, whose principal source is the sun, moves with the constant velocity of light (ie

178

3×10^8m/sec) in a harmonic wave pattern so that a reciprocal relationship exists between wave frequencies and
wavelengths. The term 'harmonic' is applied to waves
which are equally and repetitively spaced in time, and
the energy is measured in units known as *photons*. Each
part of the electromagnetic spectrum has unique energy
characteristics and wavelengths. In photography, the
major form of radiant energy which is detectable is
white light – the visible radiation which can be dis-
persed by a glass or a crystal prism into a band of
colours ranging from violet through blue, blue-green,
green, yellow, orange red and deep red within wavelengths
from about 400nm at the blue end to about 700nm at the
extreme-red end, thus forming the *visible spectrum*
(Fig 5.1). It should be noted that although this visible
spectrum occupies only a very small portion of the whole
electromagnetic spectrum, its energy is probably the most

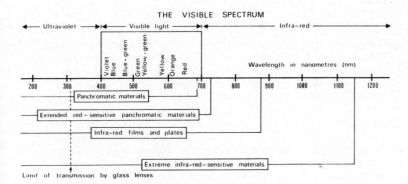

Fig 5.1 *The visible spectrum*

important as a source of information of our terrestrial
environment when compared with other new *remote sensing*

179

techniques recently developed to exploit the invisible portions of the spectrum. This is because the human eye is more accustomed to see the significance of the resultant photographic images through their tone contrast, sharpness and stereoscopic impression. Of these three characteristics, the special importance of tone contrast as a clue in photo-interpretation has already been amply demonstrated. Therefore, a gain in interpretability will result from the procurement of photography in the proper part of the visible spectrum through the effects of greater tone contrast.[1] This realisation has led in recent years to an impressive growth in the range of photographic materials for interpretation.

The following account attempts a more detailed examination of the characteristics and contributions of the three major types of special photographic systems - orthochromatic, infra-red and colour - in the extraction of qualitative and quantitative data regarding our terrestrial environment.

ORTHOCHROMATIC PHOTOGRAPHY

The orthochromatic film is a monochrome film most sensitised to light with a wavelength of 560nm in the middle of the green band of the visible spectrum. Since within its limits of sensitivity its coverage includes the shorter wavelengths at about 400nm, it is also sensitive to blue light. In the past, it was used as a standard film for aerial mapping in the United States, but this function has long been superseded by the panchromatic film. Theoretically, it should be possible to simulate the orthochromatic effect from the panchromatic emulsions by means of a minus-red filter which is capable of cutting down the long wavelength (red light) as well as the short wavelength (blue light) within the visible portion.[2]

This special sensitivity to the light in the spectrum between wavelengths of 400 and 600nm has been found particularly useful in two types of application: (1) coast studies and (2) vegetation surveys. For the former, Sonu has already pointed out that at such a range the film allows maximum depth penetration of water bodies,[3] whilst for the latter, green vegetation registers a good

tonal contrast on orthochromatic film with its strong
sensitivity to green light, thus facilitating the
identification of details about the vegetation. This
latter use is particularly favoured by the Russians who
discovered that orthochromatic photographs taken in
summer and early spring gave greater tone contrast
than panchromatic for such commercially important tree
species as pine and alnus whilst no difference was noted
in autumn between the two films.[4] Orthochromatic film
is now only used in test flights. But more recently,
the growing interest in multispectral or multiband photography has attracted more attention towards this type of
film.

BLACK-AND-WHITE INFRA-RED PHOTOGRAPHY

This utilises the conventional camera lens to focus an
infra-red ('below the red') image on to a photographic
emulsion sensitised to infra-red radiation so as to
produce a black-and-white negative record and subsequently
a positive print. This infra-red radiation is therefore
photographically actinic and lies in the range of
wavelengths from 700nm to about 1,350nm in the electromagnetic spectrum. There is another type of infra-red
radiation with wavelengths longer than about 1,350nm
which exists as heat patterns and can only be imaged by
non-photographic means.

  Because of the wide range of wavelengths covered, the
infra-red emulsions are sensitive to violet, blue and
red light of the visible spectrum in addition to infra-
red. A minus-blue (ie yellow) or a red filter of 600nm
(eg Wratten 25A) or of 680nm (eg Wratten 89A) has to
be employed over the camera lens at the time of photography to eliminate these undesirable light rays so
that only the infra-red radiation is recorded. The
energy so recorded may be transmitted, reflected and/or
emitted from the object according to its atomic and
molecular structure. It should be noted that, owing
to the longer wavelength of infra-red compared with the
visible light, it focuses further from the film plane,
thus requiring the visible focus to be adjusted. Modern
aerial camera lenses such as the Wild Universal Aviogon
lens and the Zeiss (Oberkochen) Pleogon A lens have been
so well designed that they are chromatically corrected

for the infra-red part as well as the visible part of the spectrum and are therefore free of these troubles.

Certain features of infra-red films are particularly outstanding, including the following:[5]

1 Special reflection properties of vegetation caused by the fact that the epidermis and pigments of the leaf are very transparent to infra-red radiation, thus allowing it free access to the leaf mesophyll (including both palisade and spongy parenchyma cells) from which the near infra-red radiation is reflected and scattered.[6]

2 Better haze penetration ability which is made possible by the film's sensitivity to radiation of much longer wavelengths so that the scattering effect of the dust particles in the atmosphere is very much reduced.

3 A high degree of absorption of infra-red light by water bodies rendering water features pitch-black on infra-red film.

4 Black shadow effect which is due to the insensitivity of the infra-red emulsion to diffuse or polarised light.

All these characteristics can be usefully exploited either singly or in combination in the following fields of application: (1) plant ecology, (2) soil and hydrology, (3) geology, and (4) glacial geomorphology.

1 *Plant Ecology*. The fact that the infra-red radiation is reflected mainly by the tissues of the leaves, irrespective of the surface colour, has been utilised to distinguish between different kinds of plant. Thus, the broadleaf and needleleaf trees can easily be differentiated as the amount of reflectance from the former is much higher than that from the latter, especially in the infra-red wave band (Fig 5.2), but finer species identification within each group is not possible as they record to about the same tones.[7] Another application is to study the health condition of the plants. When the plant life is under stress from its environment, the leaf mesophyll of the plant may be affected. When this happens, the plant generally loses its infra-red reflectance, a fact which can be recorded by a change towards much darker tones of the vegetation than usual on the film. The causes of the stress are usually diseases and insects, an early detection of which will

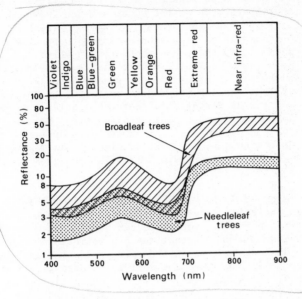

Fig 5.2 *Spectral diagram for foliage reflectance of broadleaf and needleleaf trees (after Colwell, 1965)*

save great losses of our plant resources. The film therefore also assists an understanding of plant pathology.[8] Based on these facts, one can determine the different growth conditions of plants and see whether these are ideal or submarginal.

An important point to note about the application of infra-red photography in plant ecology is the effect of climatic and seasonal variations. Infra-red film is mostly used in the temperate latitudes in summer when poor differentiation occurs in the green band.[9] But the good haze penetration quality of the film means that it is also useful in the humid tropics where atmospheric haze poses a serious problem in aerial photography. There have also been experiments in the use of different filters (other than the 'normal' minus-blue or red filters) in combination with the film to counteract certain disadvantages of infra-red photography such as the occurrence of black shadows which obscure features. Absorption filters such as the Kodak Wratten 12 which transmits green light (500nm) and Wratten 15 which transmits orange light have been employed to produce the so-called 'modified infra-red photography' which is particularly favoured in the United States. The modified infra-red results in more luminous shadow detail, and gives excellent results for mixed coniferous-broadleaved forests and for differentiation between species by tone within the conifers and broadleaved forests. More frequently, panchromatic film is used in conjunction with infra-red to give more information. This may be regarded as the simplest type of multispectral photography.

2 *Soils and Hydrology*. The fact that water absorbs infra-red radiation to a high degree gives rise to applications where the soil type can be evaluated on the basis of its moisture content, as the wettest soil should register darkest in tone on the infra-red with minus-blue photography.[10] But where the soil is covered with vegetation, the tonal change will not correctly reflect the moisture content of the soil. Colwell quoted an example where the swampy spots in meadows appear lighter in tone on infra-red photographs than do the dry spots because these swampy areas are covered with grass leaves.[11]

This property of infra-red photography leads to applications in hydrological research, especially in (a) drainage analysis and (b) coastline delineation. The exact course of small streams can easily be recognised, especially if they run through a forested area where the contrast between the water features and trees comes out most sharply so that even small and hidden jungle streams can be detected with ease (Plate 22). In the quantitative analysis of drainage networks, the value of infra-red photography has been demonstrated by Parry and Turner.[12] They made use of modified infra-red photography obtained with a Wild RC-8 aerial camera and the Kodak Infra-red Aerographic film type 5424 with a Wratten 12 filter at a scale of 1:16,868. This was then compared with panchromatic photography (Kodak Plus X Aerographic type 2401 exposed with a Wratten 12 filter) at a similar scale (1:17,367) over the same forested area of New Brunswick, eastern Canada. The drainage networks were traced on the overlays monoscopically by examining the respective panchromatic and infra-red photographs under a magnifier and their lengths measured using a travelling micrometer. The results clearly revealed the superiority of infra-red photography in improving channel detection and identification by approximately 37 per cent compared with panchromatic photographs, and showed that it was best suited for detecting second- and third-order channels with progressively less advantage for larger ones with streams ordered according to Strahler's system. Even for the finest first-order streams, infra-red photography showed an average performance level of 73 per cent compared with only 39 per cent for panchromatic. In addition, shallow channels were revealed through bar complexes, and the extent of shallow water inundations could also be accurately indicated.

This latter aspect leads to the second use of infra-red photography, ie in shoreline studies. It should be noted that only clear water is photographed very dark on infra-red film, whilst muddy water shows up in lighter tones. This fact may be employed to detect sources of polluted water. Thus, infra-red photography has been employed to delineate the land-water contact of tidal country as demonstrated by Jones[13] and Theurer.[14] In a study of the Dyfi Estuary, on the coast of Cardigan Bay, Jones observed that the use of standard infra-red photography was limited in coastal studies by its inability to penetrate water and advocated the use of infra-red photography with a minus-blue filter as this combination would produce photographs with the power of depth penetration whilst still giving a sharp waterline.[15] He also noted the ability of infra-red photography to expose the old river channels. In the United States tide-controlled infra-red photography has now become a standard procedure for establishing the mean high water mark and the mean low water mark in coastal mapping.[16]

3 *Geology*. Infra-red photography permits lithological differences of the terrain to be more clearly reflected in the grey tone pattern than panchromatic photography, thus facilitating the identification of rock types. Lattman drew attention to the differences in renditions of shale and limestone in infra-red photography.[17]

Similarly, rock outcrops underneath a not-too-dense vegetation can be more easily detected. All these depend on the level of infra-red reflectance from different rock types.

4 *Glacial Geomorphology*. Infra-red photography has been employed to study the different types of glacial depositional features, as demonstrated by Winkler[18] and Welch.[19] Welch, in particular, has compared the usefulness of infra-red photography with panchromatic and colour photography in the interpretation of these features at the Breidamerkur Glacier area in Iceland, but he concluded that the infra-red film (Kodak Infra-red Aerographic which possesses an effective range of the spectrum sensitised to wavelengths 600-900nm) was the least satisfactory for this purpose owing to the very high contrast between ice and gravels; a great loss of details in the shadow areas; a slight resolution fall-off; and, in practice, the difficulty experienced in determining the correct exposure of photography; whilst the advantage was the ease with which vegetation and water courses could be detected and traced out respectively.

It has been seen that the applications of black-and-white infra-red photography lie mainly in extracting information about the natural environment. As for applications to the built environment, not too many examples can be quoted. There has been some discussion of the power of infra-red photography to penetrate industrial haze which might be prevalent over a polluted factory zone in the city, as suggested by Maruyasu and Nishio,[20] but Welch pointed out that even in this aspect infra-red photography does not offer too much advantage over panchromatic minus-blue photography.[21] The author himself has also carried out research into the use of black-and-white infra-red photography in the extraction of qualitative and quantitative information about the urban environment, taking the city centre of Glasgow as an example.[22] The film used was Kodak Infra-red Aerographic Type 5424 which has a spectral sensitivity between 400nm and 900nm of the spectrum. It was observed from the Glasgow stereomodel that the new buildings in the city appeared more distinctly in much lighter tone than the old ones and that the different types of road surfacing came out clearly in different shades of grey. But in a densely built-up area, the greatest limitation of all appeared to be the shadows cast by the buildings which were recorded as pitch-black, obscuring valuable details that were found in them. Therefore, the infra-red photographic system is not at all ideal for use in analysis of the urban environment. Apart from this, the metric quality of infra-red photography is obviously poorer than that of panchromatic, as the setting of the floating mark on the street level would become difficult because of these shadows. Therefore, the conclusion that the infra-red photographic system is best used in the natural environment for rural land use survey and in open terrain for photogeologic study becomes inevitable.

Finally, one other related area of application of infra-red photography is in archaeology and historical geography. Strandberg has demonstrated that many crop and soil marks and other land scars which are not visible on the ground or on panchromatic photographs stand out clearly on infra-red photographs, leading to discoveries of sites of ancient settlements.[23]

Despite all these distinctive characteristics of the black-and-white infra-red photographic system, one should note that there are also limitations. One of these is its much lower speed as compared with the panchromatic. Modern developments have largely overcome this difficulty; eg for aerial photography, the commonly used Kodak Infra-red Aerographic Type 2424 has an Aerial Exposure Index of 100 which is comparable to medium-speed panchromatic film. The other limitation is its lower resolving power. For the 'Plus-X' (panchromatic) Aerographic film the resolving power is 40 lines/mm but for the Infra-red Aerographic, it is only 28 lines/mm, both at a test object contrast of 1.6:1. However, these two types of film are now available on Estar base so that good dimensional stability can be maintained. Of course, infra-red film is about 20 per cent more expensive than panchromatic film and one must determine whether the additional information obtained justifies the extra cost involved. The storage of infra-red films presents more difficulty than that of panchromatic films, as they have to be kept in a refrigerator at 14°C or colder. Also, after exposure, the films have to be processed immediately to avoid undesirable changes in the latent image.[24]

COLOUR PHOTOGRAPHY

All photography using monochrome films fails to give a true picture of our terrestrial environment in the sense that it cannot depict colour as we see it. Strandberg pointed out that many features in the environment have unique colours associated with them such as *green* grass, *yellow* corn, *blue* water, *golden* wheat and *scarlet* maples.[25] The addition of a new dimension of hue to photography therefore eliminates much of the feeling of unreality that occurs when one attempts to search for the significance of the various shades of grey during inter-pretation. It has been pointed out that as the human eye can separate 20,000 hues and chromas, but only 200 shades of grey, much more information can be obtained from colour photography than from black-and-white photo-graphy.

The colour that we see is really light of different wavelengths inside the visible spectrum (ie 400 - 750nm) which has been modified by reflectance from or trans-mittance through some substance and reaches the retina, the light-sensitive surface of the eye. The retina con-sists of a fine array of light-sensitive cells which appear in two forms: rod-shaped and cone-shaped. The rods, which are much more sensitive to light, are not sensitive to colour and can only discern objects in terms of light and shade. On the other hand, the cones can operate at higher brightness levels to sense colour. It is interesting to note that colour sensation is pro-duced by the response of only three types of receptor in the cones, one sensitive to blue light, one to green, and one to red. These receptors are affected in different proportions by different colours; and the blue, green, and red are called primary colours because they cannot be made from the other colours.

This is the principle utilised for the production of colour films which makes colour photography possible. In general, one can distinguish two basic techniques of colour sensation by our eyes: (a) the additive method, and (b) the subtractive.

The *additive method* simply mixes the three primary colours of blue, green and red in appropriate proportions.

188

It would be possible to photograph an object three times on panchromatic emulsions through a blue, green and red filter separately. Each of the negatives so produced would be a record of one primary colour in the original object; and by changing each into a positive transparency and projecting each through the appropriate filter in superimposition on a screen, the colour of the original object photographed would be reproduced. In this process, it is obvious that there is a large loss of light involved and the white colour cannot be so easily reproduced.

The *subtractive method* may be regarded as the reverse of the additive method. Instead of starting with the three primary colours, it starts with the white light, and by taking out varying proportions of the primary colours from the white light independently, a wide range of colours can be reproduced. This is achieved by using absorbing dyes having colours which are complementary to the primary colours. Thus, it was discovered that a yellow dye absorbs blue, a magenta (purple) dye absorbs green and a cyan (blue-green) dye absorbs red.

The subtractive method is superior to the additive method as it largely overcomes the difficulties experienced in the latter. Modern colour films are therefore constructed on this principle. A true colour film is made up of three layers of emulsion, one for each of the three primary colours, blue, green and red, in the order from top to base (Fig 5.3). Since the green and red layers are also sensitive to blue light, a yellow filter is inserted between the blue and green layers to eliminate the undesirable blue light. After exposure, each layer is developed out in the three complementary colours of yellow, magenta and cyan corresponding to the colour sensitivity of each layer. There are two systems of colour film processing: the colour positive or reversal system and the colour negative system. The positive or reversal process involves first of all black-and-white development, followed by re-exposure of the residual silver halide in each layer with diffuse light, and then colour development with the production of subtractive colour dyes. All the remaining silver halides and metallic silver are removed in a bleach bath followed by fixation of the image (Fig 5.4). The negative process

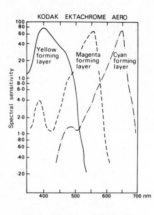

KODAK EKTACHROME AERO

BLUE SENSITIVE	YELLOW FORMING LAYER
FILTER	FILTER
GREEN SENSITIVE	MAGENTA FORMING LAYER
RED SENSITIVE	CYAN FORMING LAYER
BASE	BASE

Fig 5.3 *Sensitivity and cross-section of a true-colour film (after Jones, 1971)*

does not require the black-and-white development and re-exposure. The colour forming compounds have already been included in the respective light sensitive layers during manufacture and a single stage of colour development produces the complementary colour in each layer (Fig 5.5). There are colour films which can be processed to positives or negatives as desired, eg Kodak Ektachrome which is normally intended as a reversal film but can be pro-cessed to a negative by the Aero-Neg process. The sub-tractive method also has its limitations in practice. The major one is the difficulty in obtaining subtractive dyes having the required colour purity; and the method called colour correction masking has to be used. Usually, the magenta dye is the one which most needs colour correction.

There has been much discussion of the suitability of the two systems of colour film processing for aerial photography.[26] In general, one can say that the colour

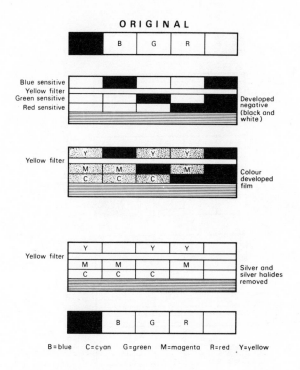

ORIGINAL

Blue sensitive
Yellow filter
Green sensitive
Red sensitive

Developed negative (black and white)

Yellow filter

Colour developed film

Yellow filter

Silver and silver halides removed

B=blue   C=cyan   G=green   M=magenta   R=red   Y=yellow

Fig 5.4 *The method of reversal colour processing*

reversal film is more versatile and economical as the
resultant positive can be directly used in photogrammetry
and photo-interpretation; but the colour negative film
gives higher resolution and allows duplicate prints to
be made, thus allowing some control over the colour during
printing. But the choice of a suitable processing system
is really connected with a multitude of problems that
arise when colour films are applied to aerial photography.
This is largely related to two requirements: (a) the
degree of reality and clarity of the terrestrial environment that can be depicted to aid photo-interpretation;

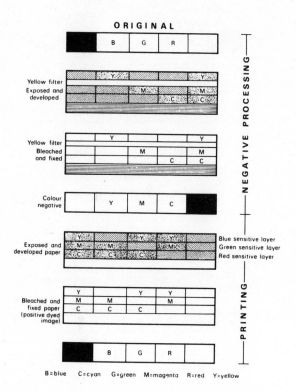

ORIGINAL

Yellow filter
Exposed and
developed

Yellow filter
Bleached
and fixed

Colour
negative

Exposed and
developed paper

Blue sensitive layer
Green sensitive layer
Red sensitive layer

Bleached and
fixed paper
(positive dyed
image)

NEGATIVE PROCESSING

PRINTING

B=blue   C=cyan   G=green   M=magenta   R=red   Y=yellow

Fig 5.5 *The method of negative colour
processing*

and (b) the degree of metric accuracy that colour aerial
photography is capable of giving from measurement or in
topographic mapping. These can be met only when the
photographic system (the aerial camera) and the materials
(the colour films) in use are correct.

For the first requirement, the aim of colour aerial
photography is to maintain the colour fidelity of the
terrestrial environment, ie to obtain what Meier called
colour-correct aerial photography.[27] It has been

192

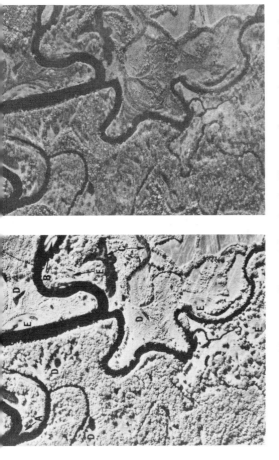

▼ *Plate 21* A panoramic photograph of Downtown Boston, Mass, showing Kenmore Square Area and Pruden-
tial Center Tower taken with an Itek 76.2mm Optical Bar Panoramic Camera from a height of 487.7m (*Itek
Corporation*)

Plate 22 Infra-red (*left*) and panchromatic (*right*) photographs of the Oromucto river area of New Brunswick, Canada. The infra-red photography was taken with a Wild RC-8 camera (f = 152.15mm) on Kodak
Infra-red Aerographic Film Type 5424 with a Wratten 12 filter. The panchromatic photography was taken
with a Zeiss RMK 15/33 camera (f = 153.04mm) on Kodak Plus-X Aerographic Film Type 2401 exposed
with a Wratten 12 filter. The original scale of both photographs is about 1:17,000. Note the superiority of the
infra-red photograph in emphasising drainage courses and minor fluvial features. The features indicated are
(A) channel bar accretion, (B) minor channel through a channel bar complex, (C) meander cut-off, (D) back
swamp, and (E) point-bar swale swamp (*Parry and Turner, Department of Geography, McGill University, Montreal*)

Plate 23 Itek's nine-lens multiband camera *(Itek)*

Plate 24 D-Mac System-2 Digi-Grid solid-state digitiser *(D-Mac Ltd)*

observed that colour film is very sensitive to ultra-
violet radiation which causes colour distortion, but
fortunately this can easily be eliminated by means of
filters, and the photogrammetric camera lenses exhibit
strongly reduced transmission in these ultraviolet wavelengths so that the effect is not at all important. But
there are other factors that tend to affect the colour
rendition of the resultant photographs. These include
(a) the nature of the light sources, (b) the reflectance
of the object, (c) atmospheric absorption and haze, and
(d) the exposure differences. There are two light
sources: the sun and the sky; and it is well known that
sunlight has a greater red component whilst sky light
contains much more blue. Thus, a bluish hue can be
observed in the shadow regions illuminated exclusively by
sky light. As the light strikes the ground, it is not
reflected uniformly in all directions by the objects, but
rather produces different reflectance angles which in turn
rigorously alter the colour stimulus of the objects. But
in the course of reflecting the light back from the
objects through the atmosphere to the camera lens, a part
of the light is absorbed whilst extraneous light (ie
atmospheric haze) is added. The colour stimulus is again
changed. The effect of haze on aerial photography is
dependent on the angular field and the flying height. In
general, as the angular field and the flying height
increase, the fall-off in illumination increases resulting
in an increase in the blue component so that apparently
the best colour aerial photography can be achieved with
narrow angular fields and low flying heights. Finally,
the exposure differences can produce changes in colour
rendition, and the aerial camera lenses do exhibit some
irregularities in light distribution in the image plane
and in a light loss of about $\cos^4 \alpha$ (where α is the angle
of light incidence to the lens) with 100 per cent loss
at the axes and 25 per cent at the corners, which can
only be counteracted by means of anti-vignetting filters.
These factors combine to make colour distortion inevitable in the resultant colour photographs, and this
requirement for colour-correct photography is extremely
difficult to meet. One needs to specify to what extent
colour distortion is permitted.

For the second requirement, the aim is to maintain a
degree of metric accuracy comparable to that of black-

and-white panchromatic film. This involves consideration
of the dimensional stability of aerial films and the
metric quality of aerial cameras. Recent developments by
the film manufacturers have made possible the use of a
stable polyester base for colour film, and even colour
glass diapositives can be produced. The main difference
between panchromatic and colour film lies in the fact
that the latter has three layers of emulsion, ie much
thicker emulsion than the single coating in the former.
After processing, this may produce larger radial distortion and hence poorer image resolution. Today, film
manufacturers such as Kodak have produced a large variety
of colour aerial films to overcome many of these difficulties. Thus, the Kodak Ektachrome Aero Film Type 8442
and Ektachrome MS Aerographic Film Type 2448 on Estar
base are respectively high-speed and medium-speed colour-
reversal films for medium-to-high and low-to-medium
altitude reconnaissance purposes. The Ektachrome Aero
has a less stable film base (being made up of cellulose
triacetate) than the Ektachrome MS Aerographic which has
a polyester base with fast-drying backing. More recently,
the Kodak Ektachrome Aero Type 8442 has been replaced by
Kodak Aerocolour negative on film type 2445 on Estar
base, which is a high-speed, extremely fine-grain, colour
negative aerial film and can be processed into a colour
negative only. It has certain outstanding properties
such as moisture-resistance, superior dimensional
stability and very high resistance to tear, and its
emulsion is abrasion-resistant, all of which are aimed
at meeting the stringent requirements of the photogrammetrists for metric accuracy; hence its stated
application to low-to-medium altitude aerial mapping.

The other consideration is the aerial camera. Apart
from satisfying the geometric requirement for preserving
a true central perspective, it is also necessary to
consider the camera's photographic power and the uniformity and spectral balance of illumination throughout its field of view. For the aerial camera to be
capable of taking metrically correct colour photographs,
the high-performance camera lens has to be chromatically
corrected for the range of spectral sensitivity of the
most-used colour films. More important is the need to
correct for the unavoidable light fall-off, especially
as the angle of view of the lens increases.[28] The use

of anti-vignetting filter is generally insufficient for
this purpose in colour photography because the exposure
latitude of colour films is much more limited than that
of black-and-white panchromatic. There is also a need
to reduce the spectral transmission of the short wavelength region. Colour correction filters have to be
used. But filters for colour aerial photography are
rather more special and have to be constructed from
pre-cut gelatin sandwiched bewteen two plane-parallel
glass plates. Unfortunately this gelatin is most susceptible to moisture changes, producing shrinkage and
distortion which give rise to a zonal decrease in resolution. All these problems have to be carefully tackled
and a suitable combination of filter-lens-film is
essential for the best result in colour photography.
Modern aerial cameras have been designed to cope with
both black-and-white and colour photography, such as the
new Wild RC-10 Universal Film Camera which can be fitted
with a wide-angle (f=152mm) or super-wide-angle (f=88.5mm)
lens. The wide-angle lens (called the Aviogon) is
chromatically corrected for the range 450-850nm whilst
the super-wide-angle lens (the Super-Aviogon) is chromatically corrected for the range 400-600nm. Both are
said to be capable of giving a low lens distortion of
less than 0.01mm.[29] On the whole, tests conducted by
the American Society of Photogrammetry have shown that
small-scale wide-angle colour photography (eg 1:36,000)
is as accurate as black-and-white panchromatic photography for aerial triangulation work and stereo-photogrammetric plotting.[30]

Therefore, it is clear that the procurement of colour
photography is a much more complex matter than that of
black-and-white panchromatic, and the choice of a
suitable film-processing system depends on the ultimate
use to which the colour photography is to be put. It
is small wonder for photogrammetric purposes that the
colour negative rather than the colour reversal system
should be used because of the former's superior colour
quality and dimensional stability, its greater exposure
latitude, its higher flexibility in duplicating colour
and black-and-white prints on paper, film or glass,
and its more tolerant and less exacting processing procedure.[31] On the other hand, photo-interpreters may
prefer the colour reversal system as the positive trans-

parencies are superior to prints for interpretation, as observed by Welch, because they give much higher resolution than prints from colour negatives.[32]

A new system of colour photography, known as the Colour Coding System, is now being developed by Technical Operation Inc for the United States Army Engineer Topographic Laboratories.[33] This aims at using a single frame of black-and-white reversal film rather than multiple frames to obtain a colour image. The whole system operates on basic light filter grating and lens theory. A transmission gating is placed at the plane of the camera when the film is exposed through a red, green and blue filter separately. Before each exposure, the grating is rotated through a fixed angle. It is possible to combine the sequential steps into a single opeartion so that a single exposure records the three primary colours at one time and each is modulated by a different filter grating combination. The colour image can be reproduced by projecting the processed positive through the red, green, and blue filters. It should be noted in such a system that the grating spacing limits the present resolution of the colour coding system to 50 lines per millimetre.

Applications of True Colour Photography

With the growing interest in the use of colour films and *true* colour photography, it is not surprising that applications are numerous and varied.[34] However, among these one can single out two areas of special importance: (1) agriculture and forestry, and (2) geomorphology and geology.

1 *Agriculture and Forestry*. For agriculture and forestry, colour photography obviously facilitates detection and identification of crop and tree types through their colour associations. This has been credited to the fact that a wide range of colours distinct in hue, value and chroma is represented and that the subdued hues can be reproduced.[35] The colour negative process has been recommended as ideal for this purpose because colour correction is possible during the course of production of a positive colour transparency to give the optimum colour contrast for interpretation. Low-

196

altitude colour photography at scales of 1:8,000 to
1:12,000 was found to be particularly valuable for intensive management purposes. Similarly, Steiner *et al*
have shown that for an agricultural area near Zürich,
Switzerland, colour photographs taken in the month of
July using Ektachrome films gave an average accuracy of
photo-identification of 30 per cent as compared with only
11 per cent with black-and-white panchromatic or infra-
red photography. But, more important, if both June and
July photographs were used together, the accuracy was
increased to about 70 per cent.[36] Another common
application with some economic significance is in the
detection of diseased plants, because plants under stress
exhibit unique discolorations of their foliage on colour
aerial films. Wert and Roettgering have illustrated an
efficient approach by which probability sampling techniques were combined with aerial colour photography to
estimate the damage caused by the Douglas-fir beetle
epidemic in California.[37] The sampling design was the
stratified two-stage cluster sampling, which was arrived
at after examining a reconnaissance flight of the survey
area. Colour aerial photographs in stereo triplets were
then obtained by the K-17 aerial camera (focal length =
305mm) with Anscochrome D/200 colour film over the sample
locations at a scale of 1:8,000. The positive colour
transparencies were interpreted over a desk-type fluorescent light table and through an Old Delft scanning
stereoscope. All groups or individual dead trees killed
by the beetle were detected. The results revealed that
such large-scale colour photographs were capable of
giving very high correlations (0.97) of ground-to-photo
tree counts. In another application, the effect of Gamma
radiation as a form of radioactive contamination to
vegetation was also investigated, using panchromatic,
non-panchromatic and colour aerial photography.[38]

Closely related to agriculture and forestry are applications to soil surveys. The identification of soil
types is rendered easier by colour. Areas of little or
no soil development stand out very distinctly in colour
photography. Colour photography is best suited for a
non-forested area where the colour variations (or colour
contrasts) allow soil boundaries to be drawn easily. A
recent investigation has shown that true colour photography using a reversal system, the Kodak Ektachrome Aero

Film Type 8442, can best distinguish soils with high chroma (ie high colour), such as a fine sandy loam.[39]

One point that emerges from all these applications of colour photography to forestry and agriculture is that the season for aerial photography is frequently more important than the choice of film, scale or any other factor. The application by Steiner *et al* has already stressed this fact.[40] Jones, in discussing vegetation surveys, has also pointed out that in a study over the Dartmoor area the limit of bracken (*Pteridium aquilinum*) was not well defined on colour photography taken in late June, but stood out well in colour photography taken in August and December.[41] In soil studies, Simakova has observed, using the Caspian lowland as an example, that the best time for aerial photography should coincide with the season in which the various soil/plant areas are reproduced with a high degree of colour contrast on the resulting aerial photographs; and he believed that the best time for that particular area was in summer and early autumn.[42]

2 *Geology and Geomorphology*. Another major area of application of true colour photography is in geology and geomorphology. It has been pointed out by Verstappen that photogeological work in areas of scanty vegetation, particularly where igneous or metamorphic rocks occur, is greatly facilitated by the use of colour.[43] The rock outcrops and structural details can be more easily identified. Scattered outcrops and isolated occurrences of a stratigraphic unit can be detected by means of changes in the colour sequence of the various formations. Any sharp breaks in the colour sequence can also be interpreted as shear zones and other structural features (eg faults), whilst zones of alteration and mineralisation can easily be recognised by the more intense coloration typical of the minerals.[44] These advantages point towards colour photography as a useful tool in natural resources exploration, such as petroleum and minerals, as confirmed by a recent paper.[45] It is noteworthy that colour photography helped stratigraphic mapping in an area of sedimentary rock very well and became even more useful as the geology became more complicated. Therefore, the potential petroleum structures in gently folded regions could be mapped with ease. In mineral exploration, all

fault intersections and igneous intrusive contact zones which are potential sources of minerals showed well in colour photographs. One further advantages of colour photography is its good interpretability of the shadow areas which appear blue and not so dense as to obscure details. A quantitative approach along the same line of tone measurement in black-and-white panchromatic photography was indicated by Fischer who made use of the colour densitometer to measure light transmitted through colour transparencies, thus giving spectral data which could be compared with colorimeter measurements of light reflected from the actual rock specimens.[46]

   The power of colour photography to penetrate water and still permit a clear delineation of its limits has been an asset in shoreline mapping, since much more detail is visible on the sea-bed to allow more rapid and accurate identification of points. The US Coast and Geodetic Survey has made use of this in combination with infra-red photography for nautical charting purposes.[47] In addition, as colour photography can penetrate water only to a limited depth, it can be used to isolate shallow areas which may require a dense hydrographic study, whilst tide-synchronised photography is best obtained with colour films.[48] More recent applications of colour photography in coastal studies place stress on high-altitude photography, taken from a height as great as 18.3km, as a kind of simulated satellite photography, which is really only suitable for macroscopic-scale studies. But an interesting use of colour photography was that different types of coastal water could be differentiated on the basis of colour changes alone, or, more technically, the variations in the optical densities of the positive colour transparencies. Thus, along the Oregon coast in the United States, the colder and more saline upwelled water and the warmer Columbia river plume registered in different colours on Kodak Ektachrome film type 2448, and their boundaries or fronts could be clearly demarcated.[49] The location of these frontal areas was economically significant because commercially important species of fish such as tuna and salmon tended to aggregate. The use of a colour densitometer succeeded in identifying three types of water off the Oregon coast: (a) *green water* indicative of high chlorophyll and phytoplankton concentration in upwelled waters or

plume waters of high productivity near the shore; (b)
deep-blue water associated with clear offshore water
containing less chlorophyll further off; and (c) *blue
water* which was seldom found within 64.4km of the coast
and occurred more continuously off southern than northern
Oregon. The same principle can also be applied to study
pollution of the sea and river water - the so-called
aerial water-quality reconnaissance system.[50] Another
application was to use high-altitude colour photography
to study coastal processes on a macroscopic scale. Dolan
and Vincent reported the investigation of the crescentic
features of sandy coasts, in particular the so-called
sand waves and *shoreline rhythms* of fields of *en echelon*
points and embayment.[51] Photographs of 1:60,000 scale
were used; but by combining these with low-altitude
photographs (say 1:10,000 to 1:20,000 scale), a finer
analysis of the hierarchical nesting of these crescentic
forms became possible. More recently, experiments were
carried out by Eastman Kodak to improve the water-
penetrating ability of colour film so that maximum
information about underwater detail and water characteristics could be obtained from high-altitude aerial photography. These led to the development of a two-layer
film which has peak sensitivities at about 480 and 550nm
and is capable of providing maximum penetration of water
with varying amounts of organic matter present. The
top layer is blue-green sensitive and the bottom layer
is green whilst the dyes formed are respectively the
complementary colours: magenta and green. The film
is processed to a positive to maximise colour contrast.[52]

Apart from surveying the coastal and underwater environment with colour photography, other geomorphological
features such as those found in glaciated areas have also
been studied. From this, Welch, who conducted such an
investigation in Iceland, found that colour photography
offered the best all-round interpretability for the
identification of water features (lakes, ponds, stream
courses), bedrock, buried ice in kames and eskers and
different sources of rock materials.[53] The landform
patterns became more evident through the contrast of
water, vegetation and gravels. Welch attributed the
superiority of colour photography for the interpretation
of glaciated areas to the better contrast produced as
a result of the combined effect of colour (chromaticity)

and brightness, whereas in monochrome photography contrast is due to brightness only. Herein lies the major
advantage of colour aerial photography. Jones also came
to the same conclusion in his study of the Dyfi estuary
in Wales.[54]

3 *Other Applications.* Colour photography also finds
application in other areas of interest, though probably
not to the same extent as in forestry-agriculture and
geology-geomorphology, eg highway engineering, archaeology, land uses, etc.[55] It is perhaps worth mentioning
that colour photography is in fact quite suitable for use
in urban studies, especially in examining the layout and
main features of the built environment. Even at small
scales, such as 1:47,000, fine detail can still be
detected because of the 'realistic' rendering of colours;
and as a result, urban and industrial uses can easily be
delineated.[56] At larger scales, such as 1:10,000 and
1:15,000, colour photography allows even more detailed
study of the urban environment, in particular, the
detection and delineation of urban slums or urban poverty
areas.[57] This application should be particularly useful
in studying the Third World cities where slums, squatters,
and other forms of poor housing are usual. However, from
an examination of the bibliography prepared by Manji[58]
and Kracht and Howard,[59] it is disappointing to see that
applications of colour photography to the human environment very much lag behind those to the natural environment, and further researches in this direction are
certainly needed.

FALSE-COLOUR PHOTOGRAPHY

It has been mentioned previously that the maintenance of
colour fidelity is one of the major requirements of good
colour photography, and indeed in some applications the
'realistic' colour rendering is deemed a major advantage.
But there is one type of colour photography which de-
liberately distorts the true colour of the objects in
order to achieve better interpretability through even
greater colour contrast. This is false-colour photography, which may be regarded as a type of colour photography using a special type of colour film. In essence,
this colour film achieves the false-colour effect by
means of a 'colour shift'. Instead of being sensitised

to the three primary colours of blue, green and red, the
three layers of emulsion of false-colour film are made
sensitised to green, red and infra-red separately. A
yellow (or minus-blue) filter is used at the time of
photography to cut out blue light. Upon development,
positive images of yellow, magenta and cyan record in the
green-, red- and infra-red-sensitive layers respectively.
This arrangement is best illustrated by an example of
Kodak false-colour film which is also known as the false-
colour infra-red: Kodak Ektachrome Infra-red Aero film
type 8443 (Fig 5.6). This type of film was originally

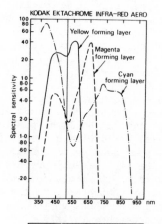

Fig 5.6 *Sensitivity and cross-section
of a false-colour film (after Jones, 1971)*

designed for camouflage detection purposes and hence was
known as camouflage detection film; a Wratten filter
number 12 had to be used. Depending on the proportions
of green, red and infra-red radiations reflected or
transmitted by the objects, numerous colours which are
'unrealistic' of the objects are formed. It should be
noted that the film does not react to radiation emitted

by terrestrial features. Therefore, false-colour film may be regarded as a colour counterpart of black-and-white infra-red film. Kodak has recently replaced Ektachrome Infra-red Aero Film Type 8443 with Aerochrome Infra-red film Type 2443 on Estar base - a reversal film with much higher dimensional stability and better keeping qualities. Another type is Aerochrome Infra-red Film Type 3443 on Estar thin base which is a reversal film for forest survey and camouflage detection and replaces Ektachrome Infra-red Aerial Film Type SO-180. It is noteworthy that no negative system for false-colour film is available from Kodak because it cannot be satisfactorily developed as a negative.

The advantage of the false-colour system lies in its modified colour rendition of the subject; because the visible light component is added to the infra-red record the film produces characteristics colours in photographs of many physiological and botanical subtances. It is the infra-red component that produces the modified colour. A list of the subjects and their associated colours is shown in Table 5.1. It is obvious that false-colour photography is best applied to practical forestry and agricultural land-use studies. A detailed discussion of its advantages in forestry is given by Stellingwerf, who applied it to study trees in the temperate latitudes.[60] One of these advantages is the fact that healthy trees can be separated from diseased trees through colour differentiation, even though crown symptoms of the disease may not be visible. This is distinctly superior to true-colour film which can differentiate between healthy and diseased trees only where crown symptoms of the disease are visible. Healthy broadleaved species, which reflect mainly in the infra-red range and to a small extent in the green range, are, with transmitted light, photographed as red. On the other hand, conifers, which have less infra-red and more green remission, give, with transmitted light, an image containing less red and more blue. However, Benson and Sims, working in Australia, strongly queried the ability of false-colour film to detect early or 'previsual' symptoms of disease of trees, and tended to support the use of true-colour photography for damage and mortality determination.[61] Colwell did point out that previsual symptoms of many but not all plant diseases were

Table 5.1 False-Colour Renditions of Certain Subjects As Produced by Kodak Colour Infra-red Film

	Subject	Colour
1	Healthy, deciduous, green foliage	red, magenta
2	Diseased or deficient foliage	greenish, bluish
3	Badly stressed foliage	yellow
4	Conifers	dark purple
5	Evergreens	red brown
6	Red rose	yellow
7	Blue flowers	yellow
8	Blue sky	sky blue
9	Some green dyes	magenta
10	Some green pigments	purple
11	Some black cloth	dark red
12	Dolomite-limestone	grey-brown
13	Khaki cloth	orange-red

Source: *Kodak Infra-red Films*, Kodak Publication
        No N-17 (nd), p4.

discernible on photography sensitised to infra-red wave-length of about 700 to 900nm.[62] Hildebrandt and Kennweg, working in Germany, confirmed that different colour renditions occurred in accordance with the different condition of the trees.[63] Thus, they discovered that (a) defoliated trees always appeared in a blue or green colour; (b) dead but not defoliated trees did not always appear blue or green, but may be greyish straw colour or greenish grey, depending on the time when the tree was killed; and (c) injured trees which still have a completely and dominantly live foliage never appear blue or green, but will vary ranging from yellow through grey and brown to pink inclusive of all possible transition colours. This difference in idea may be due to the fact that in Australia the dominant vegetation is the eucalypt which has little or no spongy mesophyll tissue in its leaves and is therefore not sensitive to infra-red radiation as are other vegetation types in temperate areas. More recently, Wiegand *et al* have found that the optical behaviour of leaves in the 0.75-1.35μm near-infra-red band is one of high reflectance and low absorptance

204

affected by leaf mesophyll structure, whilst the 1.35-2.5μm band is influenced more by the optical properties of water in the tissue than by the leaf structure. They therefore concluded that vegetation discrimination and stress detection were most effective in spectral bands centred at 0.57, 0.65, 0.68, 0.85, 1.65, 2.0, and 2.1 or 2.2μm.[64] Another recent piece of work in the use of aerial photographs for detection of bark beetle infestations of spruce in Sweden has shown the importance of the time factor.[65] Its authors found that one could detect attacks 7-14 days old and that the changes in the trees showed up faster during the later periods of attacks. The general conclusion, however, was that colour firms were superior to black-and-white and that the colour infra-red film was better than normal colour film for the detection of infested trees.

Despite these complications, the use of false-colour photography has become well-established in forestry and in agricultural land-use studies. In the Canadian Forestry Service, medium-scale (1:4,000) colour infra-red aerial photographs were solely employed for classifying forest damage into four basic categories based on morphological or physiological variations exhibited by the trees.[66] In the United States, Rhode and Olson reported its use in detecting trees suffering from disruption of water metabolism caused by insect and disease attacks. The foliage of such trees usually develops higher moisture tensions (high leaf temperatures) which were recorded as purple colour on colour infra-red film.[67] Another application along the same lines is the wetland mapping project carried out in New Jersey involving the use of both true- and false-colour aerial photography at the scale of 1:12,000. An inventory was carried out at the same time. The wetlands are located along the marine coastal zone and tidally-influenced estuaries which are distinguishable by characteristic botanical associations. The final maps are in two scales, 1:2,400 and 1:6,000, and each contains the upper wetland boundary, the line of biological mean high water, and the delineation of major plant species associations, all of which were interpreted from the false-colour photographs. Field checking by helicopter was also carried out and confirmed that accuracy was generally sufficient to meet the standards of the National Ocean Survey in the United States.

Altogether twenty-one maps at 1:2,400 scale were produced for the area in 105 days.[68] A more quantitative approach to forest studies has been developed by Turner who measured the optical density values of the colour infra-red positive transparencies obtained at regular time intervals by means of a micro-densitometer.[69] This was to study the spatial and temporal changes in the volumes of green plant material. The major problem was to standardise these measurements, as the photographic conditions tended to vary. To achieve this, densitometer measurements were made with the red, green, blue and visual filters. The colour densities obtained with the red, green and blue filters were expressed as percentages of the densities obtained with the visual filter. It was found that the red image was particularly useful as an indicator of changes in plant volumes and conditions.

As for agricultural land-use studies, work carried out in the Isle of Man proved the superiority of false-colour photography to other types of photography as applied to an analysis of moorland and woodland vegetation and an examination of field land-use.[70] Rough grass, dormant bracken, growing gorse and areas recently burnt as well as individual tree species in mixed woodlands and plantations could all be identified with ease. In the cultivated lowland, even different types of pasture, individual crops and crop blight could be identified from colour infra-red photography. A more thorough evaluation of false-colour photography in rural land-use analysis has been carried out by Samol for the Asheville Basin test site in western North Carolina.[71] Both small- and large-scale wide-angle photography, respectively, at 1:30,000 and 1:7,000, were used. Altogether three types of film, panchromatic, Ektachrome true colour and Ektachrome colour infra-red, were compared. It was found that of those landscape features fundamental to the interpretation of land use, the false-colour film was particularly well-suited to record three: terrain, drainage and vegetation. This was credited to the infra-red component which enhanced shadows, increased the con-trast of the water with the surrounding materials and gave distinctive vegetative signatures in colours. In brief, all the characteristics of black-and-white infra-red were further amplified in colour infra-red. In actually interpreting the agricultural land use, the film

206

was particularly suitable in bringing out the farmstead morphology through strong contrast with its surroundings, the detection of which gave criteria to the identification of farming types as discussed above under black-and-white panchromatic photography in Chapter IV. Also, the boundaries of the farms, movement patterns, crops and crop-systems, farming techniques and soil erosion could all be interpreted with ease. It was further pointed out that the probability of significant features being overlooked during interpretation was less with the use of colour infra-red although the false-colour gave an 'unrealistic' view of the feature which had to be 'translated' before maximum information could be derived from the film. Finally, colour infra-red photography was found to exhibit less reduction in colour differentiation accompanying scale decrease than true-colour photography as a result of its stronger colour contrast and better haze penetration. This means that smaller-scale aerial photography is equally useful, thus allowing interpretation of macroscopic features such as topography, transport and drainage networks, and regional distribution of agricultural activities. Therefore, the conclusion was that false-colour photography was the most versatile and highly valuable of all the photographic systems, a view shared also by Cooke and Harris. On the other hand, Jones, whilst realising the great potentiality of false-colour infra-red film, was less certain of its more widespread use because its film base stability was poor.[72] The new Kodak Aerochrome Infra-red Film Type 2443 on Estar base that replaces the old Ektachrome Infra-red Type 8443 and exhibits less variation in emulsion quality may have overcome these metric deficiencies. Further tests on its use are certainly required.

So far the applications of false-colour infra-red photography have been confined to forestry and agricultural land uses. Certainly, other applications outside these two areas are possible. In fact, false-colour photography can be applied to all those areas where true-colour photography has been applied, and invariably it tends to produce better results. Thus, geological and soil mapping can benefit from the stronger colour contrasts and a wide range of colour variations afforded by colour infra-red film.[73] In particular,

colour infra-red is best used to distinguish soils with
a low chroma (ie soil grey or neutral in colour) such as
a silty clay loam.[74] In livestock studies, sheep counting
was best done with false-colour photography, which also
accentuated the differences between over-grazed and
healthy ungrazed vegetation as an indication of grazing
pressure in the area.[75] Its use in highway engineering
is equally suitable as it can point out the sources of
building materials even more sharply. Water pollution
analysis can be more effectively carried out with false-
colour infra-red aerial photography which images the
sulphuric acid contaminated water and soil areas in cyan
hues.[76]

   As for applications to the human environment, the major
concern is still with urban areas. There are three areas
of interest in which colour infra-red photography has
proved its superiority. Firstly, in urban land-use
surveying and mapping where rather detailed information
is usually required, colour infra-red photography was
reported to be capable of emphasising characteristics of
many cultural features, such as building outlines, roof
design, road materials and conditions, and ground surface
conditions, even more than true-colour photography.[77]
Secondly, in housing quality studies, colour infra-red
photography at the scale of 1:6,000 was again found to
be most useful in estimating seven variables (surrogates)
important in housing quality identification, which were
(a) on-street parking, (b) street width, (c) street grade,
(d) hazards from traffic, (e) access to buildings, (f)
refuse, and (g) roading and parking hazards.[78] From this,
one would see the suitability of this film in delineating
urban poverty areas and slum areas. Thirdly, in dwelling
unit estimation from aerial photographs where each
housing type really needs to be identified according to
the external morphology of the building, such as roof
type, shape, structure, size, height, division of buildings, and associated features, such as the availability
of parking, presence or absence of vegetation in the area,
etc, Lindgren found that colour infra-red photography was
particularly suitable for application to high-density
residential areas because details appeared much sharper.[79]
Also, a much smaller photographic scale could be employed
with this type of film than with panchromatic or true-
colour films. In fact, the scale of 1:20,000 was used

for colour infra-red photography over the Boston metropolitan area and an accuracy of 99.5 per cent in the
identification of residential structures was achieved,
which compared favourably with those obtained by Green
(99.8 per cent) and Binsell (99.9 per cent) both using
black-and-white panchromatic photography at the scales
of 1:7,500 and 1:4,800 respectively.[80]

MULTISPECTRAL PHOTOGRAPHY

From the foregoing review of the applications of different
types of non-panchromatic and colour photography it has
become increasingly clear that no single film-type is
ideal for all purposes. One type of film may exhibit
some advantages in certain aspects, but limitations in
others. Although true-colour film has been regarded as
being capable of more general use, it suffers too much
from the necessity to maintain correct colour fidelity
which is highly elusive depending on how the weather
changes. If cost is unimportant, the combined use of all
the different film types will certainly yield more information than any one of these as our terrestrial
environmental facets are so highly varied.

It has already been mentioned that infra-red photographs when used in conjunction with panchromatic photographs can yield more information than when each is used
alone. This fact has also been stressed by Colwell, who
demonstrated the complementary nature of these two types
of photography in rural land-use interpretation.[81] For
example, on panchromatic and infra-red aerial photographs
of a rural area in the Sierra Nevada foothills of
California taken on the same date, he noted the superiority of the panchromatic photography in discerning dry
stream channels, field boundaries, vegetation boundaries
and roads and, on the other hand, the superiority of the
infra-red photography in discerning moist stream channels,
vegetation boundaries, wet soil and rocky hummocks in the
same area. This is the underlying principle of 'multispectral' or 'multiband' photography.

Multispectral photography involves sensing simultaneously different portions of the electromagnetic spectrum
(known as spectral regions) in order to accumulate more
information. This exploits the fact that 'transmission,

reflection, absorption, emission and scattering of
electromagnetic energy by any kind of matter are selective with regard to wavelength and are specific for that
particular kind of matter, depending primarily upon its
atomic and molecular structure'.[82] This means that each
object has its own 'spectral signature'. In practice, a
multiband camera is employed by means of which several
photographic film-filter combinations, each specially
suited to sensing in its own spectral band, are used
to obtain photographs of the same area at the same time.
Both the multi-camera and the multi-lens designs are
possible. The multi-camera design involves the use of
several identical cameras each with its own lens and film
whilst the multi-lens design makes use of only one camera
body with several lenses and one or more rolls of film.
The multi-camera type is more popular from the geographical point of view because of its flexibility in film-
filter combinations and its lower cost; and only small
frame (70mm) cameras need be used. There are four-,
six- or nine-lens cameras; and usually, for the nine-
lens type, photographs are obtained with panchromatic
film for six of the wavebands and with infra-red film for
three. One important example of the multispectral photographic system in use today is the nine-lens camera
produced by the company of Itek in the United States
(Plate 23). This is really a multi-lens design, with all
the nine lenses mounted on a single lens cone, each lens
covering a 70mm format. The camera magazine contains
three spools of 70mm film on each of which are projected
three images. This is more flexible than the other
multi-lens design in that each channel can be used with
a different type of film if greater flexibility in
spectral sensitivity is desired, and spectral filters can
also be placed in each of the nine lenses to obtain
simultaneous recording of a scene in nine bands. Each
is a 152mm f/2.8 Schneider Xenotar lens chromatically
corrected for the band in use. Image motion compensation
is provided. A further advantage is that the camera can
be fitted with a magazine containing a 23cm film (ie
standard format) to obtain at nine bands recording
separately on one photograph.[83]

A thorough evaluation of this complex technique of data
acquisition has been reported by Pestrong who applied it
to study the tidal marsh in the vicinity of San Francisco

Bay, California.[84] The nine-lens multiband camera of
Itek was used to obtain the photographs, and the nine
spectral bands covered were: 400-440nm, 460-500nm, 525-
550nm; 550-590nm; 600-630nm, 650-710nm, 700nm, 775nm; and
825nm. In addition, panchromatic (Kodak Plus-X), colour
(Kodak Ektachrome Aero) and colour infra-red (Kodak
Ektachrome Infra-red Aero Film, Type 8443) were obtained,
so that in fact twelve 70mm-frame photographs were
produced simultaneously for photo-interpretation. Subjective as well as objective evaluations of the photographs were carried out. The former involved a tracing
analysis whereby the drainage networks, water penetra-
bility, vegetation differences and environmental variation
were directly traced from each of the twelve transparencies on the light table. The latter required the use
of a microdensitometer which measured the density across
each transparency in a series of traverses. Pestrong
discovered that (1) the near-infra-red photography (ie
bands 7, 8 and 9) gave the highest degree of precision
in the delineation of drainage patterns; (2) the 550-
630nm (ie orange and red; bands 4 and 5) region of the
visible spectrum permitted the greatest degree of water
penetration and hence gave the greatest clarity of water
bottom phenomena; (3) the 550-710nm bands (ie bands 4,
5, and 6, or green-yellow, orange, and red) were very
useful for distinguishing among zones of plant varieties,
and also gave the greatest clarity in distinguishing
between the different types of tideland environments
based on tide levels. He also pointed out the superiority of colour infra-red photography for the differentiation of vegetative types and delineations within the
marsh environment, whilst colour photography was good
for general interpretation. The potential of the microdensitometer as an interpretative tool was stressed and
its use in the automation of photo-interpretation was
hinted at. One particularly important conclusion from
this evaluation was that the nine-lens multiband imagery
was excessive with a great deal of duplication. Pestrong
suggested a more economical and equally efficient photographic system constructed of four synchronised cameras
utilising colour, colour infra-red, near-infra-red and
the 550-630nm band (ie green-yellow and orange) of the
visible spectrum.

A four-camera photographic system was adopted by Yost

and Wenderoth for applications to oceanographic and agricultural studies.[85] This system contains four 177.80mm focal length f/2.5 lenses matched in both focal length and distortion. Either a single piece of film for all four images or four individual pieces of film can be used so that a high degree of flexibility in film-filter combination is possible. The image motion compensation is also incorporated. The idea of this multi-spectral system is to produce four photographs corresponding to the three primary colours (the blue band (395-510nm), the green band (480-590nm) and the red band (581-715nm)) and the infra-red band (700-900nm). The sets of four spectral negatives are carefully developed so that the density of the image on each individual negative is a correct representation of the brightness of the object; and a correct gamma (ie the slope of the density-exposure curve) should be chosen to produce good contrast. From these negatives, positive transparencies are obtained and are superimposed in good register in a multi-spectral viewer for interpretation. Thus, a colour composite is created in which the density differences between positives appear as a colour; and where no density difference occurs, the image is only a shade of grey. This makes possible the detection of subtle density differences. The multispectral viewer has the further advantage of controlling the colour brightness of the image by means of a filter and a saturation lamp. In other words, by means of this system, true-colour, false-colour and spectrozonal photographs may be produced as one wishes. This technique has been successfully applied to agricultural studies involving the identification of tree species, crops and soil types as well as the different growth conditions of the plants, using small-scale (1:44,000) photography.

As for oceanographic applications, the characteristics of different coastal waters were studied to see how these affected the detection of submerged objects. High-speed film (Kodak 2485) has to be used and it was necessary before the multispectral photography to measure the reflectance of the submerged objects and hence to determine the types of filter used for the photography. The multispectral photographs obtained were then combined to be viewed through these filters inside the multispectral viewer as a spectrozonal image. Thus, it was found that as the transmission wavelength of the filters approaches that of maximum water transmission, the visibility of underwater objects improves. Also it was observed that as the water gets less and less clear, the primary transmission wavelength from the object gets longer. More recently, Helgeson also made use of the nine-lens camera to study the performance of the films in determining water depth and came to the conclusion that the ideal range was from 450nm to 500nm.[86]

An important new development stimulated by the work of Yost and Wenderoth in superimposing four multiband colours into a colour composite image for efficient viewing in a multispectral viewer was the single-lens multiband camera designed by the Engineer Topographic Laboratories in the United States which involves the use of three dichroic beam splitters and four magazines so that four photographs recording the blue, green, red and infra-red energy reflected from the terrain can be obtained at one exposure (Fig 5.7).[87] This unique design allows perfect

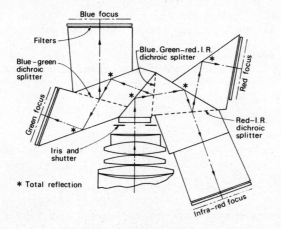

Fig 5.7 *Beam splitter single lens multiband camera system developed by the US Army Engineer Topographic Laboratories (after Anson, 1972)*

213

registration of the four photographs and eliminates the necessity of bore sighting for matched lenses in the conventional multi-lens design.

In general, one can see that the applications of multispectral photography fall mainly in the fields of land evaluation and resource inventory. Tanguay *et al* illustrated an application using fifteen different wavebands of the electromagnetic spectrum from 320nm to 14μm for soil mapping and evaluation for engineering purposes.[88] Both the subjective and objective approaches in the interpretation of the multispectral imagery were carried out; and it was obvious from visual comparison that the visible red band (620 to 660nm) was most useful for soil studies by bringing out the soil contrasts. On the other hand, the objective evaluation employed the microdensitometer to make a series of density maeasurements on the multispectral imagery which were then normalised against a standard grey scale so that the tones of the imagery recorded within different spectral regions were compatible. The potentiality of the microdensitometer as an automatic data classifying tool was confirmed. Orr and Quick similarly made use of the multispectral photographic system to identify the likely sources of materials for engineering construction which pinpointed positively such geomorphological features as river bars and active beaches in delta areas.[89] There are also applications to wheat surveys,[90] disease detection,[91] forestry,[92] environment analysis,[93] and geological and natural resources surveys.[94]

It is evident from the recent flourishing of interest in multispectral photography that the system is the most efficient reconnaissance tool of our terrestrial environment which is capable of collecting abundant information in a short time; but this efficiency tends to lag at the interpretation stage whereby the human interpreter has to correlate visually a multitude of spectral signatures of the objects in four or more wavebands. There is also the inevitable element of subjectivity. Therefore the need for some forms of instrumental aid capable of performing more objective evaluation of multispectral imageries is obvious. This explains the importance of the microdensitometer as an image tone measurer and the multispectral viewer used by Yost and Wenderoth in

detecting subtle density differences. But the ultimate
development will be towards full automation in photo-
interpretation or image evaluation as the multispectral
photographic data continue to pour in.

CONCLUSIONS

In this chapter, the principles and uses of non-panchromatic and colour photography have been explained and
reviewed. One common characteristic of all these new
photographic systems is that they have been developed to
aid photo-interpretation and to provide more specialised
data regarding our terrestrial environment. These data
are therefore essentially qualitative in nature. This
points up two implications: (1) black-and-white panchromatic photography can satisfy very well the high accuracy
of the quantitative data extracted and the metric accuracy
of the topographic map completed; and (2) there is an
increasing need for new and more detailed qualitative
information relating to our terrestrial environment as
world population exerts pressure on food and natural
resources (ie the ecological view).

For the first point, certainly some of the photographic
systems can be used in topographic mapping. True-colour
photography has been credited with being as versatile as
panchromatic photography for aerial triangulation work
and large-scale topographic mapping. This was confirmed
by tests conducted by official mapping agencies such as
the United States Geological Survey[95] and the Ordnance
Survey in Britain,[96] and by private institutions such as
the American Society of Photogrammetry[97] and a commercial mapping company.[98] The added new dimension of
colour allows better interpretation of detail to be
carried out for topographic mapping purposes. Indeed,
more accurate height measurement may be obtained from
the colour stereomodel. Reeves has shown that colour is
easier to read and provides a firmer surface for spot
elevation readings, thus reducing fatigue and enabling
a less experienced operator to work more quickly. The
production time, as a result, was cut by 5-10 per cent
on a full-scale mapping job. True-colour photography
was also used in tree-height measurement.[99] Even within
the built environment, building heights could be
accurately obtained, as a test conducted by the author

215

using a colour stereomodel of Glasgow city centre has shown.[100] This was due to better contrast between the floating mark and the object measured, which allowed more precise setting of the mark. Another advantage of true-colour photography was that the shadows were only dark bluish in colour and did not obscure detail. Today, colour films are much more stable dimensionally. The negative colour system was generally superior for photogrammetric purposes. Despite these advantages, colour photographs are still not popular in topographic mapping, mainly because of much higher costs in film and processing especially where duplicate prints are required. The resultant accuracy only equals that of black-and-white panchromatic and therefore does not make this higher cost worth while.

For the second point, the quest for more qualitative information appears to be far more important than that for maps. This is because maps once drawn can exist for a long time and only require up-dating from time to time. Such map revision in turn asks for more qualitative information only because metric information, such as contours, seldom changes rapidly, if at all. The increasing complexity of our built environment coupled with the need to search for more natural resources and to maintain the balance of our ecosystem have given rise to problems that can only be wisely tackled if one is properly informed. This inevitably leads to the increasing adoption of multispectral photographic techniques, development of non-photographic remote sensors tapping the invisible portions of the electromagnetic spectrum and finally automation in information extraction, storage and retrieval with the aid of the electronic computer as the cross-referencing of information becomes too complex for a human being to handle efficiently. All these will result in a systems approach and the integration of aerial photography into a much broader but rapidly expanding system of remote sensing.

VI AUTOMATION IN PHOTOGRAMMETRY AND PHOTO-INTERPRETATION

By now it is clear that the conventional aerial photographic system has greatly extended its applications and
usefulness to geography through a more diversified approach. This has been brought about in recent years by
the availability of highly advanced camera optics, which
make possible the manufacture of distortion-free lenses
of high resolving power; improved film types sensitised
to different wavelengths of light in the visible spectrum;
and the variety of camera-carrying platforms which range
from the low-altitude helicopter hovering over a small
area to the high-altitude artificial satellite revolving
round the globe. Parallel to such advances in the conventional photographic systems has been the emergence of
the non-photographic remote sensing systems exploiting
the much wider invisible portion of the electromagnetic
spectrum. These systems have proved their usefulness
in collecting qualitative data of our terrestrial environment which nicely complement those obtained from conventional aerial photographs. The result is that the
conventional aerial photographic system becomes incorporated into a much larger remote sensing system with
greatly expanded capabilities.

All these developments have inevitably given rise to
the problem of data load. The magnitude of such a problem will be better appreciated if we realise that a
single aerial photograph contains a large amount of
information depending on its scale and the nature of
the terrain covered. It has been observed by Haralick
that some planned high-resolution imaging systems are
expected to collect by themselves over 3×10^9 bits* of

---

*The term 'bit' is contracted from 'binary digit', which
is made up of Os and 1s in computer usage.

information per day.[1] Earlier, Barrett estimated that about 100,000 prints per week were produced from the aerial reconnaissance system for military purposes in the USA, and about 12,000 photographs sent back to earth from the satellites per month.[2] The figures have been greatly increased since then, especially with the Earth Resources Technology Satellites (ERTS) now circling round the earth. The ERTS system by itself is capable of providing 10,000 photographs each week covering 2,590 million hectares of the earth's surface and of relaying 70 million bits of data each day.

This greatly increased volume of photography is difficult to handle manually, and the task of extracting the qualitative as well as quantitative information from it is formidable. For example, one is required in the case of multiband photography to compare tonal values of the photographs area by area and feature by feature to obtain the tonal signatures of individual objects. This can become very complex indeed.

But behind all this flood of photographic imagery, there is an ever-increasing need for information from various disciplines; and above all, there is a real need for *accurate*, *up-to-date* and *detailed* information regarding our terrestrial environment to be supplied *quickly*.[3] It appears that aerial photography and other remote sensing systems have possessed the potential to satisfy such a need. The trends towards automation in photogrammetry and photo-interpretation only represent the first step made in this direction.

AUTOMATION IN PHOTOGRAMMETRY

The extraction of metric information from aerial photographs involves the use of photogrammetry. The data extracted are mainly quantitative measurements of the terrain, which are utilised for the construction of topographic maps. The data extracted can also be non-topographic in nature, for example, parameters relating to the forms of various cultural features, or a count of a certain type of object observed in the photograph. It has already been shown in Chapter II that both the analytical and analogue approaches can be adopted for this purpose of mensuration and map marking, and stereo-

218

plotting machines of varying orders of accuracy constructed on these principles have been available for human operation. Here, one can see the necessity for a skilled operator to control this small-scale photogrammetric system. Automation in photogrammetry aims at the minimum human participation in the photogrammetric system in order to speed up the map production process and to achieve more accuracy in the resulting map. Thus, automation in photogrammetry is basically a technical revolution and an improvement on machine systems.

In a sense, the concept of automation has been prevalent in photogrammetry for a long time. The analogue plotting machines were usually regarded as automatic machines in the past because they are a means of avoiding the tedious measurement of coordinates and the laborious computations required by the analytical approach when a stereocomparator is used. The instantaneous production of a map in 'real time' (ie plotting at the speed at which the stereomodel is being measured) by the analogue photogrammetric machines (which can be regarded as analogue computers) was believed to have exhibited a high degree of automation. It is small wonder that these machines usually bear names with automatic implications, such as Stereoautograph, Autograph, Autokartograph, etc.[4] Today, automation in photogrammetry can be regarded as the continuing development towards the perfection of these analogue stereoplotting machines to achieve full automation. This in essence means the further automation of three phases of operation of the machines, namely, measurement, orientation and recording. It is only with the advances in electronic and computer technology that the automation of these photogrammetric systems is possible. Thus, one can distinguish four main lines of development: (1) photogrammetric digitising of graphical outputs; (2) analytical or computer-controlled stereoplotters; (3) orthophotography; and (4) automatic image correlation.

Photogrammetric Digitising of Graphical Outputs

When a stereoplotter is employed to produce a topographic map, a stereopair of vertical aerial photographs is set up and then orientated to reconstruct the correct stereomodel which is then measured stereoscopically by means of a floating mark. The output is a graphical plot of planimetric details and contours, which is then passed on to the cartographers for working towards a finished map. In addition, it is also necessary occasionally to obtain coordinated positions and heights of individual points during such operations as aerial triangulation and longitudinal and cross-profile measurement along specific lines and engineering work. For these latter operations, the manual measurement and recording of X-, Y- and Z-coordinates of points for a great number of stereomodels can be very tedious and are most susceptible to human error. It is desirable that these operations can be automated. Photogrammetric digitising is therefore the means whereby photogrammetric data of the terrain are measured and recorded in a digital form for input to data-processing systems which are usually connected to a computer.[5]

The photogrammetric digitising method makes use of the principles of the *rotary digitiser* or *shaft encoder*, in which the linear X-, Y- and Z-motions are converted to a rotary motion to allow the measurement to be made. As explained in Chapter II, a high-precision stereo-plotter equipped with carriages and lead screws is connected to a very accurate coordinatograph by means of very precisely manufactured spindles which transfer the linear X-, Y- and Z-movements made in the stereomodel to the plotting arm of the coordinatograph and the height counter of the machine. The spindles responsible for X- and Y-movements are controlled by two handwheels whilst a footwheel is used to control the Z-movement of another spindle. A drum which carries a patterned metal disc is attached to each of the lead screws for these spindles so that the number of revolutions and part of a revolution of each spindle give the appropriate coordinate value (up to a precision of ±0.01mm). The drum is the rotary digitiser. To automate the process of recording the coordinate measurements, the metal disc of the drum is duplicated by a second disc of metal or glass carrying a pattern of patches which is also mounted on the shaft. The pattern is designed to generate binary code in 0s and 1s, and is arranged in such a way that different numerical values will occur at different angular positions of the disc (Fig 6.1). In the case of a metal disc, these values are monitored by a row of brushes, each of which

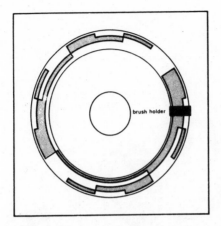

Fig 6.1 *Metal disc used in a rotary
digitiser (Rank Precision, formerly
Hilger-Watts)*

is connected to a wire. If the brush is in contact
with a metal patch of the disc, it gives an electrical
signal; if the brush rests on an insulated patch of
the disc, it will give no signal. If glass is used
for the disc, photo cells replace the brushes and the
disc is illuminated to generate the binary code. An
example of such a system is shown in Fig 6.2. These
electrical signals can be converted to give the measured
coordinate values in three different forms: (a) by direct
visual display; (b) by electric typewritten output; and
(c) by direct recording on punched cards or paper tapes.
The third form is most useful as it allows the direct
input of the coordinate values into the computer for
further processing. Also, this allows the input of more
information by means of a data entry console to accom-
pany the measured coordinate values of points. In fact,
all three forms of coordinate output can be combined in
one system, such as the Wild EK8 Coordinate Printer, an
electrical-mechanical system, which can be connected
to a large range of Wild stereoplotters: A7, A8, A9, A10,
A40 and B8. Other peripheral output devices such as the
IBM731 electric typewriter, the Wild SL15 Tape Punch,
or the IBM545 card punch can also be connected with the
EK8.[6]

Rotary Incremental Measuring System

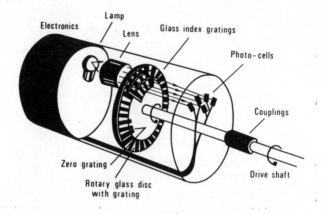

Fig 6.2 *A rotary incremental measuring system of Heidenhain, West Germany, for digitiser (after Petrie, 1972)*

It should be realised that photogrammetric digitising is mainly aimed at the automatic recording of the positions of points, and, with the use of the rather heavy coordinatograph, the system is not particularly suitable to line digitising. This is because the speed of measurement and recording for this system is relatively slow (the range of times needed to measure and record the coordinates of a single point being from 4.5 to 12 seconds) as a result partly of the time required to measure and input the coordinates into the buffer memory of the control unit and partly of the time taken to re-code and record the values on the slow output devices (such as the tape or card punch). In a thorough investigation of photogrammetric digitising as data input for processing, Petrie has shown that the rotary digitiser is not suited for use in all types of photogrammetric plotting machines, especially those equipped with free-moving carriages (eg the Galileo-Santoni Stereo-

simplex IIc) and those employing the flat-bed type of
surface as the datum surface for the measuring device
(eg the Kelsh plotter, Kern PG-2, Wild B8).[7] Even in the
type of machine equipped with lead screws and carriages,
the use of a rotary digitiser is not an advantage because
of the need to engage the shafts or carriages continuously with the encoders in order to avoid losing the respective count, thus limiting the speed of scanning and
measuring the stereomodel. Petrie therefore advocated
the use of *linear digitisers* for machines of the lead
screws type as well as of the free-moving carriage type.
The linear digitisers make use of linear gratings which
do not require shafts for their operation, and the X-,
Y- and Z-motions made in the stereomodel can be directly
measured (eg the British Ferranti-type moiré fringe and
the West German Heidenhain LIDA System) (Fig 6.3). Thus,
they can work much faster, especially in the measurement
of individual points. These linear digitisers are
particularly well suited to machines of the free-moving
carriage type, which is much less expensive than those
of the lead screws and carriages type. As for the flat-
bed measuring surface type of stereoplotting machines,
the *area, mat- or grid-type digitiser* has been suggested.
This utilises a close grid of X- and Y-position lines
embedded in a sheet of dimensionally stable transparent
material (such as epoxy resin, fibreglass or polyester)
and a sensing stylus to generate electrical impulses to
indicate positions (eg the Rand Tablet, Bendix Datagrid
or Ferranti Freescan). This can work very fast because
the linear X-, Y- and Z-motions in the stereomodel are
directly measured. More recently, the Bendix Data grid
has increased its area of coverage (up to 1.5x1.5m) and
in resolution (25μm).[8] Thus, this type of digitiser
is particularly suited to topographic plotters such as
the Wild B8 and Kern PG2 which have a flat-bed measuring
surface. These are relatively cheap machines which
geographers can afford to own for their research involving the more advanced use of aerial photography.
This type of digitiser possesses a further advantage
for geographical users in that it can be removed from
the photogrammetric machine and used over a light table
to measure data from existing maps.

The photogrammetric digitising discussed so far is
directed mainly at point digitising, but in considering

the further automation of the map production process, it is desirable to convert the original graphical output directly into digital form expressable in terms of X-, Y- and Z-coordinates for storage into the computer processing system; and an automatic coordinatograph can be used to plot the map out at different scales as required. This will of course eliminate the necessity for the cartographer to digitise the completed map manually

Linear Incremental Measuring System

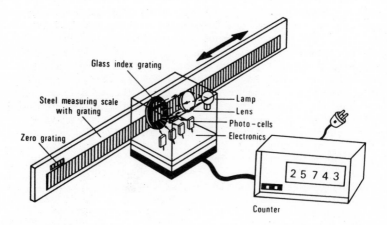

Fig 6.3 *A linear incremental measuring system of Heidenhain, West Germany, for digitiser (after Petrie, 1972)*

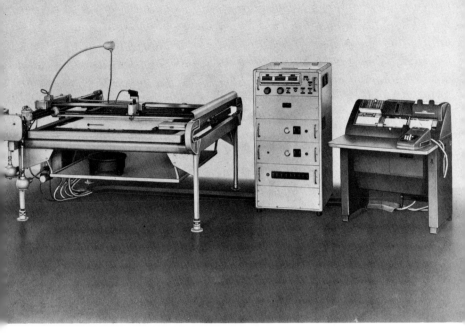

Plate 25 Zeiss Coordimat Card-Controlled Plotter *(Carl Zeiss, Oberkochen)*

Plate 26 Kongsberg 1215 Drafting Table *(Kongsberg Vapenfabrikk, Norway)*

Plate 27 Analytical Plotter AP/C-3 with IBM 1130 System *(Ottico Meccanica Italiana, Italy)*

Plate 28 Gigas Zeiss GZ-1 Orthoprojector with LG-1 Scanning Unit *(Carl Zeiss, Oberkochen)*

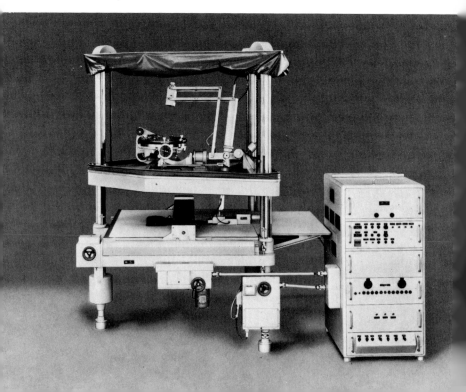

afterwards. While this *point mode of recording* of the
digitising system is suitable for recording the positions
of individual points and straight lines, its low speed
of measuring and recording is not suitable to cope with
curved lines, for which the *continuous mode of recording*
should be adopted. In this continuous mode of recording,
the measuring system is interrogated at a fixed rate
(in number of times per second) which can be pre-set on
the control panel to suit the operator's plotting speed
and the rate of recording of the output device. Thus,
the speed of recording which is determined by the nature
of the terrain in the stereomodel can be varied.

Finally, there is the *incremental mode* of recording,
by which the actual recording is only made when a change
of increment (which can be pre-set) is detected relative
to the previous recording. Thus, once the operator slows
down or stops when more difficult terrain features are
encountered, recording will also stop. This is the major
difference between this mode and the continuous mode.
As each of the three modes of recording has its own
virtues in different circumstances, these should all be
incorporated in the recording equipment for photogrammetric digitising, especially if digitising of lines is
to be done.

It should be noted that there is a major distinction
between photogrammetric digitising and cartographic
digitising. The latter is aimed mainly at digitising
the whole map, thus generating a much larger number of
coordinates. Also, it does not require the use of a
coordinatograph as in the case of photogrammetric
digitising. Hence, it is far more flexible in handling
different types of map feature. The cartographic
digitising system therefore invariably employs the continuous mode of recording. A notable example is the very
successful manually controlled type of cartographic
digitiser with which map digitising can be carried out
quite quickly, such as the d-Mac Pencil Follower produced
in Glasgow, and the Dell Foster Graphic Quantizer in
the United States. This type of cartographic digitiser
(eg the d-Mac System-2 digitiser shown in Plate 24)
consists of three basic units: a writing surface, an
electronic pen and the associated electronic equipment.
The writing surface is a vinyl-covered aluminium table

on which the map to be digitised is placed. Below the
table is a detector head mounted on a trolley which can
move along a carriage in the X-direction whilst the
carriage itself can move in the Y-direction. The electronic pen is used to follow lines on the map, and as
it does so, its stylus point produces signals in a
magnetic field generated by an AC field coil, which are
picked up by the sensor coils of the detector head and
transmitted by wire to servo motors (Fig 6.4). These

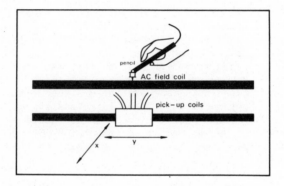

Fig 6.4 *Tracing stylus and detector
head of d-Mac Pencil Follower*

drive the trolley and the carriage so that the detector
head is always accurately positioned below the stylus
point. The coordinate positions are obtained by reading
off the shaft rotations in the X- and Y-directions by a
shaft encoder or rotary digitiser as employed in photogrammetric machines. The binary coded signals can be
converted at the attached electronic console for visual
display if required or directly punched on tapes as a
permanent record. Obviously, the accuracy with which
these positions are measured depends on the tracing
speed. A standard error of ±0.2mm is claimed for a
tracing speed of 2.5cm/sec; but it rises to ±1.0mm at
a speed of 10cm/sec (ie tracing four times as fast).[9]

Photogrammetric and cartographic digitising are not

mutually exclusive, however, and can easily be combined for more efficient functioning as a general digitising system. Thus, in the US Soil Conservation Service, a digitising system consisting of both a Wild B8 and a Dell Foster Graphic Quantizer with incremental magnetic tape recording is used.[10] In the Ordnance Survey, a d-Mac recording unit with magnetic tape recorder which can be attached to either a Wild A8 or a d-Mac CF Cartographic Digitiser was employed at an early stage of the digital mapping project.[11]

The digitising phase is obviously most essential for a fully automatic mapping system to become feasible. Automatic coordinatographs and flat-bed plotters have been developed to make use of the digitised positional information recorded on the output tapes from the digitising system. These data are fed via a punched tape reader into a control unit linked by cables to the coordinatograph, which activates the sero motors to drive the pencil (the plotting head) mounted on the rectangular X-Y cross-slide system of the coordinatograph to plot individual points and draw lines between them. Examples are the Zeiss Oberkochen Coordimat (Plate 25) which can be used only to plot points and straight lines, and the Haag-Streit/Contraves, and Coradi or Kongsberg Kingmatic device (Plate 26) which have more elaborate control equipment to allow plotting of curves. Alternatively, if a lower degree of accuracy is tolerable, the cheaper and faster computer-controlled graphical drum plotter of the incremental type, such as the Calcomp Plotter (Fig 6.5), can be used to produce maps as has been done by the Ordnance Survey at an early stage of its experimentation with automatic cartography.[12]

Analytical or Computer-Controlled Stereoplotters

The second line of development in the automation of the photogrammetric mapping process is really a revival of an old idea closely associated with the basic principle of photogrammetry which establishes a relationship between the coordinates of points as measured on the photograph and the coordinates of the corresponding points as measured on the ground. Indeed, the development of measuring and plotting instruments employing the method of stereophotogrammetry started with the

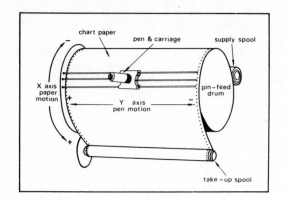

Fig 6.5 *A graph plotter of the incremental drum type*

stereocomparators designed by Pulfrich of Zeiss, Jena, in 1900 and also at about the same time by Fourcade in South Africa.[13] The stereocomparator is a machine for measuring simultaneously the coordinates of corresponding images on a stereopair of photographs. These photo-coordinates can only be converted to the correct terres-trial coordinates after much tedious computation and are then manually plotted afterwards. This is the analytical approach. Because of all these computations and the inconvenient method of plotting involved, this approach was not particularly favoured in the past and the analo-gue machines were preferred.

Ever since the early 1950s, electronic computers have been developing rapidly and have become more generally available to users. The analytical approach has been taken up again. Large national military mapping agencies have gained access to fast, large computers, for example, the UNIVAC used by the American Army Map Service, and the ACE and DEUCE machines of the National Physical Laboratory employed by the British Ordnance Survey. It becomes possible now to compute with ease the relative orientation and the terrain coordinates of a few points from photo-coordinates measured by a stereocomparator, first for a single pair, then for a strip and eventually

for a block of photography for control purposes. With
the development of advanced systems in digitising, recording and automatic plotting by coordinatographs, a new
generation of stereocomparator following the same principles of the classical Pulfrich and Fourcade stereocomparators has been developed, which is capable of much
higher precision in measuring photo-coordinates and is
equipped with automatic data-recording devices. Examples
of these modern stereocomparators designed according to
modern automatic techniques include the Rank Precision
(formerly Hilger and Watts) Recording Stereocomparator of
Britain, the Zeiss Oberkochen PSK Precision Stereocomparator of West Germany, the OMI-Nistri TA 3-P Stereocomparator of Italy, the Wild STK 1 Stereocomparator of
Switzerland and the Zeiss Jena Stecometer of East Germany
so that nearly every manufacturer has produced a modern
stereocomparator of importance.[14]

Since 1962 the advance in computer technology has
reached the stage where a small but high-speed digital
computer of suitable capacity can be constructed to
solve the problems of computing the correct positions
from photo-coordinates of a single pair in 'real time'
(ie at the speed at which they are being measured), thus
making possible the development of computer-controlled
stereoplotters. The most notable example is the
Analytical Plotter devised by U. Helava at the National
Research Council of Canada with the support of the United
States Air Force.[15] This plotter may be viewed really
as a stereocomparator coupled to purpose-built computers
and a plotting coordinatograph. The optical and mechanical components were manufactured by OMI in Italy and
the computers were built by Packard-Bell and Bendix in
the United States so that the whole analytical plotter
represents an international effort of cooperation. The
operation of this machine requires the initial input into
the computer of the calibration data (such as focal
length, lens distortion, etc) and the coordinates of
ground control points, then the measurement of the photo-
coordinates at the standard positions on the stereomodel
to determine the relative and absolute orientation.
After this, no matter which point is measured on the
stereomodel, the correct terrain coordinates will be
determined by the computer, recorded and simultaneously
plotted on a coordinatograph.

With an analytical approach, this plotter can cope with a wide variety of photographs, both conventional and non-conventional. This is why the development of this machine has received military support, and indeed the machine is too expensive for civil uses. The major application of the Analytical Plotter is to make maps from: (a) reconnaissance photography acquired by ultra-long-focus cameras, eg with f=121.9cm or 152.4cm, from 24,384m to 30,480m, from aircraft such as the Lockheed U-2 or from even greater altitudes from orbiting reconnaissance satellites; and (b) photography obtained by panoramic cameras which utilise a long focal length and produce horizon-to-horizon coverage in a single exposure.

However, a less expensive commercial version of the Analytical Plotter did come out. This is the OMI-Nistri Analytical Stereoplotter model AP-C which can handle virtually any type of photography, including narrow-angle, normal-angle, wide-angle, super-wide-angle, focal lengths from zero to 1,200mm, and vertical, tilted, convergent, oblique or panoramic photography. In 1968, the model AP/C-2 was built and in 1972 the model AP/C-3 (Plate 27) was introduced with a greatly enlarged computer storage capacity.[16] The system configuration is shown in Fig 6.6.[17] A new analytical plotter that also emerged in 1972 was the Galileo Digital Stereocartograph (DS) designed by G. Inghilleri of Italy, which comprised the Galileo MS stereocomparator, the servo-driven plotting table from the Stereocartograph V and a general-purpose Montedel Laben 70 minicomputer.[18] This growth of interest in automating the analytical approach in photogrammetric mapping reflects the mounting importance of remote sensing systems and space photography for civil applications. Recently, the emergence of relatively inexpensive mini-computers has allowed an analytical approach in topographic mapping even with the conventional stereoplotter. Dorrer and Kurz of the University of New Brunswick in Canada have demonstrated a system comprising the Wang 700A programmable desk calculator coupled to a Wild A-10 stereoplotter, which can acquire data on-line through digitising, storing, displaying and processing photogrammetric model coordinates so that relative and absolute orientation as well as coordinate transformation can be carried out among other varied operations.[19]

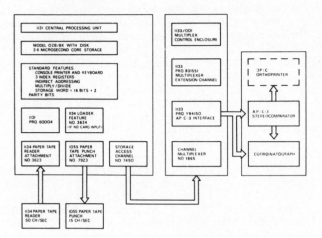

Fig 6.6 *Configuration of the OMI
AP/C-3 system*

Orthophotography

The idea of orthophotography is again an old one but has
been recently rediscovered and redeveloped. In Chapter
II it was shown that vertical aerial photography is
essentially a central projection in which image displacements and scale distortions due to the relief of the
terrain occur. This is further complicated by the errors
due to tilts, which, if suitable planimetric controls
are available, can be eliminated, using single photographs, by the process of *rectification* employing an
instrument known as a *rectifier*. In contrast, an
orthophotograph is a photographic reproduction, prepared
from the perspective photograph, in which the displacements of images due to *tilts* and *relief* have all been
removed. This means that the central projection of the
original aerial photograph has been changed into the
orthographic projection of a map, and the resultant

231

orthophotograph possesses a uniform scale and exhibits the same metric properties as those of a map. This transformation can be achieved by a method known as *differential rectification* which requires the use of stereophotogrammetry. In essence, this involves the breaking down of the terrain into a number of flat areas at different heights above the datum, each of which is then rectified.[20] In order to do so, it is necessary to know the heights of the terrain, which of course can be readily obtained by the stereoplotter.

This concept of rectification of different height zones of the stereomodel is not new, and was investigated by Lacmann in Germany and Ferber[21] in France in the early 1930s. However, their ideas were not further developed at that time. It was not until 1954 that Bean, then of the US Geological Survey, revived interest in this subject by introducing a new machine called the *Orthophotoscope*.[22] This was a successful solution to the problem which attracted wide interest and was followed by similar developments in other countries such as the Slit-Rectifier FT-Schtsch in Russia in 1959; the Orthoprojector GZ-1 of Zeiss Oberkochen in West Germany in 1964; the 693 Unit of SFOM in France in 1965; the Orthophot of Zeiss Jena in East Germany in 1965; the Orthoprinter of OMI in Italy and the Zeiss Oberkochen Ortho-3-Projector in 1967; the Wild PPO-8 Orthophotoscope and the Orthophoto-simplex of Officine Galileo in 1970; and the Kelsh K-320 Orthoscan in 1972. Most of these machines have been designed to be used in conjunction with a particular type of stereoplotter either as an independent system or, most commonly, as an attachment. Thus, the Orthophotoscope by Bean is basically a modified Balplex stereoplotter with three projectors in which the stereomodel is created by means of anaglyphic viewing (Fig 6.7). Similarly, the Russian Slit-Rectifier FT-Schtsch is a modified three-projector Multiplex. The SFOM 693 Unit has been designed for use with any anaglyphic stereoplotter. The GZ-1 Ortho-projector of Zeiss Oberkochen can in principle be attached to any high-precision plotter, but in practice it is usually coupled to the Zeiss Universal plotter - the Stereoplanigraph C8 or the new precision plotter - the Planimat. Initially, the Orthophot of Zeiss Jena was developed in conjunction with the universal mapping system - the Stereotrigomat - but more recently it has

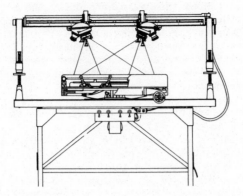

Fig 6.7 *Bean's first Orthophotoscope (1954)*

been extended for use with a much cheaper topographic
plotter - the Topocart B. The Swiss Wild PPO-8 Orthophotoscope and the Italian OMI Orthoprinter are similar
attachments to the Wild A8 Autograph and the expensive
OMI-Bendix Analytical Plotter respectively,[23] while the
other leading Italian firm of Galileo has recently
developed its Orthophotosimplex attachment for use with
its Stereosimplex line of plotting machines.[24] The Zeiss
Oberkochen Ortho-3-Projector has been developed as a
third projector for attachment to its less expensive DP-1
Double Projector (which plots maps using anaglyphic
viewing) to produce orthophotographs. The Kelsh K-320
Orthoscan is in itself a modified Kelsh plotter consisting of three projectors designed specifically for the
production of orthophotographs with much greater ease
than the orthophotoscope.[25] Thus, nearly every class of
stereoplotter ranging from the universal type to the
topographic type can be used to produce orthophotographs,
with the result that orthophotographs can be more cheaply
produced and their more widespread use becomes possible.

Despite this quite extensive and varied list, these
orthophoto-producing instruments fall neatly into two
classes according to their methods of transfer of photo-
images from the original aerial photograph:[26] (1) machines
using the method of geometrically correct optical re-

projection, such as the Orthophotoscope, the Slit-Rectifier, the SFOM 693 Unit, the GZ-1 Orthoprojector, the Ortho-3-Projector, and the Kelsh K-320 Orthoscan; (2) machines using the method of frontal (orthogonal) optical transfer, such as the Orthophot, the OMI Orthoprinter, the Wild PPO-8 Orthophotoscope and the Galileo Orthophotosimplex.

1 *The GZ-1 Orthoprojector of Zeiss Oberkochen.* For the machines in class (1), the working of the principles is best explained with reference to a specific example such as the GZ-1 Orthoprojector of Zeiss Oberkochen. The GZ-1 Orthoprojector may be regarded as one half of the Stereoplanigraph C8 Stereoplotter (Plate 28) except that the normal measuring mark is now replaced by a diaphragm with a very small slit under which is placed the photographic material required to expose the orthophotograph.[27] In producing the orthophotograph, the GZ-1 Orthoprojector is placed in a darkroom and is directly connected to the main stereoplotter outside by a mechanical or electrical transmission system in such a way that any measuring movement made in the main stereoplotter will be similarly and simultaneously transmitted to the GZ-1 Orthoprojector. For the actual procedure, a duplicate diapositive of one of the aerial photographs which form the stereomodel has to be made and inserted on the photo-carrier of the GZ-1 Ortho-projector which is then zeroed. The stereopair of aerial photographs is set up on the main stereoplotter and relative and absolute orientations of the model are carried out manually by the operator as in a normal plotting job. The tilt settings on the relevant plotting camera are then set on the orthoprojector. After these, the production of the orthophotograph begins. The main stereoplotter and the small slit of the Orthoprojector (usually 4x2mm) are driven automatically at a prede-termined uniform speed by synchromotors as a series of parallel strips along the Y-direction of the system. At the end of each strip, they are automatically stepped over in the X-direction and then continue scanning in the Y-direction and so on until the whole stereomodel is traversed (Fig 6.8). During this scanning process, the operator has to keep the measuring mark of the main stereoplotter continuously in contact with the stereo-model at the height of the terrain. When this profile

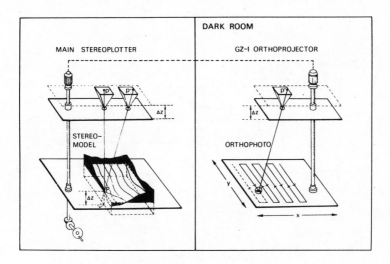

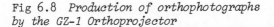

Fig 6.8 *Production of orthophotographs
by the GZ-1 Orthoprojector*

scanning process has been completed, the photographic
material can be removed and developed and the result is
a differentially rectified aerial photograph, ie the
orthophotograph at the correct scale (Plate 29).

2 *The Orthophot System of Zeiss Jena.* For the machines
in class (2), explanations are best made with reference
to the Orthophot system.[28] This consists of an optical
train whose scanning lens is directed perpendicular to
the photograph, which can move past it in a horizontal
plane only (Fig 6.9). A narrow beam of light bringing
the image from one of the photographs forming the
stereomodel is directed orthogonally towards the orthophoto plane. Automatic variation of magnification is
effected by an electro-mechanical inversor which always
ensures sharp focusing by setting up the correct distances between the image plane and the objective, and
between the objective and projection plane in accordance
with Newton's lens equation $1/u + 1/v = 1/f$ where u is

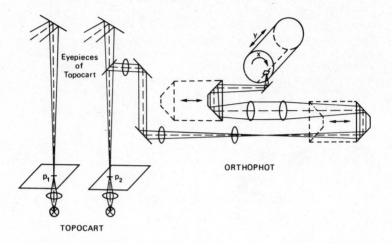

Fig 6.9 *The lens system of the Orthophot B (Zeiss, Jena)*

the image distance, v is the object distance and f is the focal length of the lens. The film for exposing the orthophotograph is wrapped round a cylindrical drum which can move back and forth in the Y-direction and can rotate to advance the film in the X-direction by an amount dx, ie the width of the strip. In this way, the scanning of the whole stereomodel is carried out using a fixed slit. The drum, the driving mechanism and the slit are all contained in a light-proof box. Thus, with the method of orthogonal projection and optical transfers of images from one of the photographs that form the stereomodel, the Orthophot eliminates the need for a third projector and a duplicate diapositive in producing the orthophotograph. One should note, however, that despite these differences in the Orthophot, the basic operation is essentially the same as in the GZ-1 Orthoprojector when coupled directly to a main stereo-plotter.

236

This Orthophot system was originally designed for combination with the Stereotrigomat mapping system, but more recently the Orthophot B was developed for on-line operation with Zeiss Jena's topographic stereoplotter - the Topocart, which is a very much less expensive machine than the Stereotrigomat (Plate 30). The Orthophot B allows the use of a variable speed up to a factor of three times during the scanning of a stereomodel, and the intensity of the exposure of the actual orthophotograph can be kept constant by using a variable-density grey wedge inserted in the optical train. The Orthophot allows differential rectification of normal, wide and super-wide-angle photographs.[29]

In both classes of machine described above, the production of orthophotographs requires continuous measurement of the stereomodel with the floating mark in a stereoplotter to which the orthophoto printing system is attached. Through this procedure of profile scanning, height information is being obtained continuously at the same instant and for the same point for which the orthophotograph is being exposed. This information can be profitably employed to produce contours. In the GZ-1 Orthoprojector, a special disc is coupled to the wheel with which the movement of the measuring floating mark is controlled. The disc has a series of different openings. During the profile scanning of the stereomodel, a light ray bundle is projected via an optical system on the disc, and, if the opening in the disc coincides at that moment to a pre-set height interval, it is exposed to form an image on a second sheet of the photographic material placed on the projection table of the Orthoprojector. The type of opening on the disc can be set to change when the floating mark passes a certain predetermined height above mean sea level. Lines of different thickness known as 'drop lines' are therefore exposed on the film, which can be used to construct contours by joining the ends of drop lines of the same types (Fig 6.10a and 6.10b).[30] In the Orthophot B, a special drop-line device called the Orograph is attached, which consists of an electronic control unit and a drawing head. During profile scanning, the Orograph can transmit a signal to the drawing head which scribes the horizontal projection of the profile lines on to a coated film sheet or glass plate placed on the drawing table at

(a)

(b)
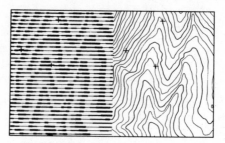

Fig 6.10 *Production of drop lines: (a)
representation of heights by using lines
of different thicknesses; (b) a portion
of the drop-line map produced with 4mm
slit width and height contours derived
from this*

the instant of passage through a predetermined contour
level. The drawing head is controlled by the elevation
of the terrain in such a manner that the line thickness
changes for every new elevation. Four different line
thicknesses are used (Fig 6.10b). Thus by superimposing
the contours obtained from the drop lines on top of the
orthophotograph, a topographic orthophotomap can be
quickly produced.

The accuracy of this differential rectification method,
by which both tilts and relief displacements of the
aerial photographs concerned can be removed, depends to
a considerable extent on the size of the slit used for
exposure. Since true orthogonal projection can be
achieved only for a point or along a profile line, the

smaller the slit-width, the greater the planimetric
accuracy of the resultant orthophotograph. The accuracy
of the differential rectification also depends on the
precision of the height values set by the operator whose
performance is affected both by the speed of scanning
and by the nature of the terrain being scanned. The use
of a slower scanning speed will allow better profiling
but a correspondingly longer time will be required to
complete the production. Also, since only the height
in the middle of the slit is set by the operator during
profile scanning, there will be a zone between the two
adjacent strips where the heights of the terrain are
wrong. Thus, by its method of production, an orthophotograph is composed of many strips whose scale was continuously changed during exposure. Mismatches will invariably occur. This becomes even more complicated if
the relief of the terrain is highly rugged. Double
images or gaps will occur depending on the trend of the
terrain slope in relation to the projection centre,
because in effect the sloping ground is being replaced
by small horizontal linear segments of the slit (Fig
6.11). Thus, it is clear from Fig 6.11 that the point
P is imaged at a higher level at P_1 in strip 1 and at
a lower level at P_2 in strip 2. This results in double
printing of the point P at the boundary between two
strips. A careful examination of the orthophotograph in
Plate 29, which was produced from the GZ-1 at the sacle
of 1:3,000 for an area around the University of Glasgow,
with a slit-speed of 2.5mm/sec, will reveal a chopping-
up effect in the roofs of buildings, bridges and roads,
especially in an area of large relief changes and near
the top and bottom edges of the photograph where relief
displacements are great. The amount of residual mismatches (dr) depends on the angular field of view from
the projection centre (α), the angle of the terrain slope
at right angles to the scanning path which is usually in
the Y-direction (βx) and the slit width which determines
the width of one strip (dx), and assumes the form:[31]

$$dr = dx.\tan \alpha.\tan \beta x$$

It follows that drop-line production will also become
more difficult in areas of extremely rugged relief; and
a lack of registration with the planimetric details
becomes usual.

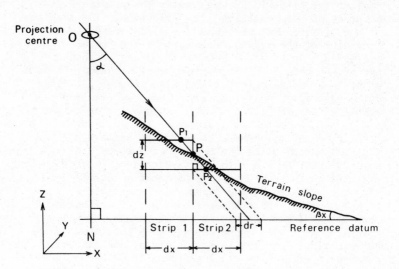

Fig 6.11 *Mismatches at the boundary between two adjacent strips caused by the terrain slope βx in relation to the projection centre O and the slit width (dx) employed in profile scanning*

The orthophotographs produced by the two optical systems described earlier have been subjected by photogrammetrists to numerous tests of accuracy in planimetry and in the heights obtained from the drop-line charts. The results of these tests showed that in general a standard error of ±0.2 to 0.3mm in planimetry and a standard error in height of ±0.4 per thousand of the flying height are achievable by orthophotographs for relatively open terrain.[32] These are considered to be quite good results.

In minimising the magnitude of relief displacement of elevated features such as buildings and trees in the resultant orthophotograph, one solution is to make use of both the left and right photographs of a stereopair to produce a single orthophotograph. This in fact

involves exposing half of the left-hand photograph near
the nadir point and half of the right-hand photograph
near the nadir point which together form the overlap.
Thus, only the best portion of the photographs where
relief displacement is least in terms of the radial distance to the nadir point has been employed in producing
the orthophotograph. This concept has been adopted by
Officine Galileo in its newly developed Orthophoto-
simplex system.[33]

It is clear that orthophotography systems possess a
very great practical potential in speeding up the
production of maps, but they are not yet fully automatic
mapping systems. Meier has observed that the most tedious
and time-consuming operation in the production of an
orthophotograph is the profile scanning stage, which
occupies 38-50 per cent of the total working time per
stereomodel.[34] This also makes great demands on the
operator who has to pay full attention in scanning the
model and is not allowed to stop or make any mistakes
before one strip is finished. Another time-consuming
phase of the whole operation is the relative and absolute
orientation of the stereomodel to be done manually in
the main stereoplotter and the orthoprojector, which
takes about 22-25 per cent of the total working time of
a model. Some recent developments have aimed at a
higher degree of automation and the elimination of the
tedious profile scanning by the human operator. An
example is the off-line production of orthophotographs
with the GZ-1 Orthoprojector made possible by the use
of a storage unit SG-1 and a scanning unit LG-1.[35] The
stereomodel is first measured using the main stereoplotter, and all the vertical profiles and orientation
marks for orientation checks on the control points are
scribed on the storage plate located in the storage
unit (Fig 6.12). Afterwards, the storage plate can be
inserted into the scanning unit in which the profiles
are scanned with the aid of the photo-cells and fed to
the GZ-1 Orthoprojector during reprojection. Thus, the
off-line operation (or the operation in the storage mode)
allows errors in profiling to be corrected before the
production of the orthophotograph and interpolation of
vertical profiles between the stored profiles to be made.

More recently, further improvements of the GZ-1 Ortho-

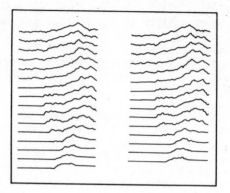

Fig 6.12 *An example of the Orthoprojector storage plate prepared in
the SG-1 Storage Unit*

projector have taken place to resolve the problems of
image mismatches and the defects of the drop-line technique. In order to increase the planimetric accuracy of
the resultant orthophotographs, Zeiss Oberkochen has
developed an 'Optical Interpolation System' which
utilises fibre-optic bundles as an image-transferring
medium.[36]

The other development for the GZ-1 Orthoprojector is
the "HLZ Electronic Contourliner' which allows the direct
production of contour lines rather than dropped lines
from the stereomodel (Fig 6.13). An off-line approach is
used since heighting accuracy can only be improved
through interpolation between profiles recorded on the
storage plate.

The off-line approach in orthophoto production is now
much favoured. Even the United States Geological Survey
has developed a new Automatic Orthophoto System to
replace its classical Orthophotoscope. This system consists of three components: the *Analog Profiler*, the
Autoliner and the *Orthophotomat*.[37] The Analog Profiler
is essentially a Kelsh Plotter modified for use in
scanning the profiles of the stereomodel at a speed

Fig 6.13 *A contour map (Waldmatt area, original scale 1:5,000) produced with the aid of the HLZ Electronic Contour-liner for the GZ-1 (Zeiss, Oberkochen)*

adjustable to the ease of the human operator. The profiles obtained are graphically plotted and are input to the Autoliner which then automatically follows the plotted profiles by means of a photoelectric cell, thus controlling the vertical movements of the Orthophotomat. The Orthophotomat is therefore a single-projector differential rectifier similar to the GZ-1 Orthoprojector with pre-set orientation elements for exposing ortho-photographs automatically.

It becomes obvious at the present state of the art that orthophotography as a mapping system suffers from the lower accuracy of the contours (with standard error about

243

0.4-0.5 per cent of the flying height) as compared with those obtained with the conventional stereoplotter (about 0.2-0.25 per cent of the flying height).[38] It is also clear that the only way to improve this is to employ a method originally proposed by Drobyshev in Russia[39] and subsequently expanded by Makarovic.[40] Drobyshev advocated the use of the conventional stereoplotter to produce a contour plot from which the height profiles would be extracted to guide the floating mark (and slit) in the subsequent traversing for orthophotograph production. In theory, the height data for each profile can be stored on tape or punched cards which are then fed to the differential rectifier or orthophotoscope to produce the orthophotograph automatically. The method of deriving profile heights from the contour plot as described by Drobyshev was a manual one and time-consuming. Makarovic therefore suggested a more automated approach whereby contours from the stereoplotter are simultaneously digitised so that the metric data of the terrain relief are converted to electrical signals to control the printing of the orthophotograph. An example under development by Wild is known as the Digital Automatic Contouring System (DACS) which makes use of the Wild EK8 Coordinate Recording System coupled to the Wild A8 with PPO-8 Orthophoto Equipment.[41]

Finally, a novel system which has been advocated by Blachut of the National Research Council of Canada involves the idea of *stereo-orthophotographs* originally proposed by Collins.[42] This involves the production of a *stereomate* for the orthophotograph so that a three-dimensional, metrically correct stereomodel of the terrain can be obtained. The stereomate is in fact an orthophotograph of the second photograph of the stereopair produced by introducing horizontal parallaxes P_x which are proportional to the elevation differences h of the terrain over a reference plane, ie

$$P_x = c.h$$

where c is a constant. Thus, the stereomate is an *oblique* projection of the stereomodel on the same reference plane as the orthophotograph which is an *orthogonal* projection (Fig 6.14 and Plate 31).

244

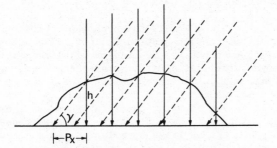

Fig 6.14 *Basic geometry of stereo-orthophotographs (after Blachut, 1972)*

To produce the stereomate simultaneously with the orthophotograph, a machine called *Orthocartograph*, specially designed by the National Research Council, has to be used. The metrically correct stereomodel is re-constructed by means of another machine called *Stereo-compiler* which is also used to extract the metric data of the terrain by measuring the X-parallaxes (P_x) and to plot contours as well as planimetric details (Plate 32). It is also possible to carry out numerical de-termination of coordinates by means of linear encoders in another version of the machine, thus allowing an analytical approach to mapping.

Experiments carried out with this stereo-orthophoto-graph system in the National Research Council have shown that even for large-scale topographic mapping (1:2,500 and 1:5,000) the standard errors of the contours came up to be 0.58 and 0.46 per mille of the flying height respectively, which are much better than those attain-able with the drop-line method.[43] Another major advan-tage is that no discrepancies exist between the ortho-photograph image and the contour plot from the Stereo-compiler. Finally, one should note that such a system requires the use of a rather simple machine to extract the metric information. The Stereocompiler is essentially a stereoscope coupled to a parallax bar and a plotting arm and can be operated with ease. Its use in photo-interpretation is also possible. Here is a system which has potential for non-topographic applications and should be of interest to geographers.

It should be noted that all the orthophoto production systems described in this section are not fully automatic. Irrespective of whether the on-line or off-line mode is being employed, the initial profile scanning stage has to be done manually. The intervention of a human operator is essential in setting up the stereomodel in the stereoplotter and in carrying out relative as well as absolute orientation. The automation of these processes involves the concept of image correlation, which leads to our fourth line of development in photogrammetry.

Automatic Image Correlation

In automatic image correlation, the human operator needed to carry out relative and absolute orientation of the stereomodel is replaced by cathode ray tubes and photo-multipliers. The essential components of a cathode ray tube are shown in Fig 6.15. It consists of a tungsten filament (similar to that in a light bulb) which heats the cathode nearby. The cathode, when heated, emits electrons which are attracted towards the anode by the electric field creating a high-voltage supply of electricity between the cathode and anode. The electrons pass through a small aperture in the anode and through another defining aperture and move on towards the fluorescent screen, which, being coated with luminescent material, will emit light when an electron beam hits it. This electron beam can be shifted to any position on the screen by means of deflection plates attached near the anode. The amount of deflection is proportional to the voltage applied across the deflection plates, and by applying a suitable deflection voltage, the beam can be made to hit the screen and produce a light spot at any desired position. Thus, by varying the deflection voltage, this light spot can be made to move across the screen so that the CRT can be used as a flying-spot scanner. As for the photo-multiplier, it is another electronic device that can convert the transmitted light into an electrical signal by means of a series of cathodes and anodes.

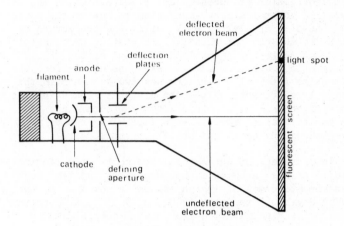

Fig 6.15 *Essential components of a cathode ray tube (CRT)*

The idea of automatic image correlation was developed by Hobrough.[44] He made use of a cathode ray tube (CRT) from which the scanning light spot emitted on the surface is projected back via the lens of a Kelsh-type optical projection plotting machine on to the diapositive. The light passing through the diapositive will show different intensities corresponding to the tonal variations of the diapositive itself. The light is first concentrated by a condenser and then measured with a photo-multiplier which converts it into a time-varying voltage. For a stereopair of aerial photographs, two photo-multipliers are used to monitor the two diapositives, thus giving two time-varying voltages. These voltages are fed into an electrical analogue computer, the so-called *correlator*, which can filter, differentiate, delay and smooth these signals. If identical signals are registered, ie the corresponding images are viewed, the correlator output is zero (ie a lack of parallax). If

247

there is a time-lag for one of the images, it shows
either that there is a lack of relative orientation
(shown by a Y-parallax signal) or that the CRT is set
at the wrong height (resulting in an X-parallax signal).
The correlator produces an error voltage which can be
used to control a servo-mechanism to remove the source
of the discrepancy, eg by rotating the projectors to
achieve the correct relative orientation or to set the
correct height by a vertical movement of the CRT. Thus,
automatic image correlation makes use of the fact that
the time interval separating any two elements in the
signal is proportional to the distance separating the
corresponding image points along the scan line; and
accurate measurement of X- or Y-parallax can be made
depending on the alignment of the stereopair.

This concept of automatic image correlation forms the
basis of most automatic photogrammetric systems such
as the Stereomat B8, which assumed a new form as the
Wild-Raytheon Stereomat A2000 in 1968[45] and, more recently,
the Zeiss Planimat with Itek EC-5 Electronic Correlator.[46]
These are the less expensive automatic systems. The
Stereomat B8 is particularly successful and had already
undergone several stages of improvement before Wild
finally stopped its production in 1970 and replaced it
with its PPO-8 and the DACS. In this system, relative
orientation is quite automatic using the concepts of
electronic image correlation as explained, but human
intervention is still required in absolute orientation
in identifying the control points (Fig 6.16). The output
is in the form of orthophotographs and drop lines. In
fact, the same profiling mode of operation is employed as
in the Orthophotoscope or Orthoprojector. However, the
orthophotograph is produced by the method of electronic
image transfer (similar to optical transfer) which requires the use of a third CRT to repeat the image from
one of the two other CRTs and to expose in blue light its
image via a lens on to the blue-sensitive emulsion of an
unexposed two-layer film. Simultaneously, a fourth CRT
is used to produce drop lines at pre-set intervals in red
light on the red-sensitive emulsion of the same film
(Fig 6.17). Also, the profile values can be automatically
recorded in digital form on punched cards, punched paper
tapes or magnetic tapes, so that the whole stereomodel
can be digitised. It is therefore clear that in this

automatic system the three major features of automation

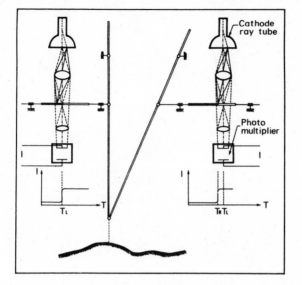

Fig 6.16 *Principle of automatic parallax detection as used in the Wild-Raytheon Stereomat*

discussed so far, namely, digitisation, orthophotography and automatic image correlation, have all been incorporated.

The other automatic system, the Zeiss Planimat with Itek EC-5 Electronic Correlator, is less versatile.[47] The stereomodel has to be set up manually on the main stereoplotter, the Planimat, in the usual way, but after that, automatic profile scanning of the stereomodel can be carried out by cathode ray tubes and correlators. Orthophotographs and drop-line plots are produced by the GZ-1 Orthoprojector either on-line or off-line via a storage unit as described earlier.

Thus, the importance of orthophotographs as an output form is seen because, as Schermerhorn has pointed out, orthophotographs offer the only means of automating the measurement of an individual stereomodel.[48] In fact, the

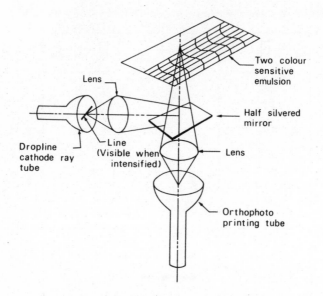

Two colour
sensitive
emulsion

Lens

Half silvered
mirror

Dropline
cathode ray
tube

Line
(Visible when
intensified)

Lens

Orthophoto
printing tube

Fig 6.17 *Simultaneous production of*
orthophotograph and contour line segments
in the Stereomat

other more expensive automatic systems such as Bunker-
Ramo's Universal Automatic Map Compilation System
(UNAMACE) or the OMI-Bendix Automatic Analytical Plotter
AS-11 B/C have also made use of this form of graphical
output.[49]   More recently, G.L. Hobrough, who originally
invented and developed the Stereomat system, has introduced an entirely new automatic orthophoto printing
system known as the Gestalt Photo Mapper, which is
essentially a correlator-equipped analytical plotter
optimised for orthophoto production.[50]   The fast Nova
mini-computer is used and the photo image is correlated,
rectified and printed out patch by patch rather than by
means of a moving slit as used in the conventional
orthophoto-producing machines described earlier.   Each
patch is a hexagon of approximately $48mm^2$.   In fact,
the printing sequence requires only a period of 5-12
minutes during which the 600-1,000 individual patches of
a stereomodel can be correlated and printed.

One should note, however, that the use of automatic image correlation is not without its difficulties as correlation may not be possible over rugged terrain where abrupt changes in height take place in a short distance, eg building heights. This is even more troublesome if automatic correlators are used in a profile scan to produce orthophotographs. The automatic system cannot differentiate between the top of an object and the ground surface. As a result the associated drop-line plot for the orthophotograph will not record the ground level. At the present state of the art of automation in photogrammetric mapping, automatic systems are still not widely used for civilian purposes because of the heavy expense involved. Perhaps one way in which other people such as geographers can make use of these automatic systems is through a bureau service which has already been started by Hobrough in Vancouver, Canada, in connection with their Gestalt Photo Mapper.

A new development in image correlation that emerged at the 12th International Congress of Photogrammetry held in Ottawa in 1972 was the use of the principle of epipolar scan correlation in the production of orthophotographs. The epipolar plane contains the perspective centre(s), two image points (a, b) and the corresponding terrain points (A, B) (Fig 6.18). Epipolar scanning and correlation involve finding the conjugate (ie the corresponding) epipolar lines on the pair of photographs and sampling the densities found along them. In doing so, the correlation problem is reduced from two dimensions to one dimension because conjugate imagery can always be found along conjugate epipolar lines regardless of terrain elevation and photograph orientation.[51] The American firm of Bendix is now developing an epipolar-scan automatic stereomapper based on certain components of the AS-11B analytical plotter specifically designed for generating digital terrain models on tape (ie numerical representation of the terrain surface) which can then be subsequently employed to interpolate contours and to produce other forms of cartography. Masry has also reported a similar application by modifying the AP/2C Analytical Plotter interfaced with the IBM 360/50 computer, in which image densities are converted to digital form and stored together with the coordinates on a magnetic tape.[52] Such developments have already

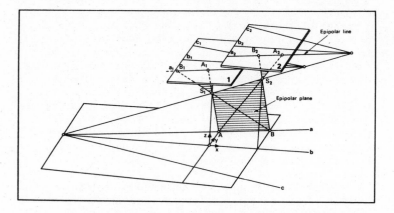

Fig 6.18 *Idealised epipolar geometry*
(after Helava and Chapelle, 1972)

attracted attention and will solve some of the problems
of automatic image correlation, especially the difficulty
caused by sharp relief changes, discussed above.

AUTOMATION IN PHOTO-INTERPRETATION

The extraction of qualitative or descriptive information
from aerial photographs requires the use of the techniques of photo-interpretation. The purposes are to
identify the objects and to judge their significance.
Doing so, as Lueder has pointed out, involves a *deductive*
and *inductive* evaluation of the various elements detected
on the photograph in terms of common sense and field
experience, supported by the interpreter's academic and
practical background.[53] In other words, human decision
and judgement have to be exercised.

Although photogrammetry and photo-interpretation have
been separately dealt with here, one should note that it

is in fact difficult to draw a clear-cut boundary between the two, because in using the photogrammetric method of map plotting, photo-interpretation also comes in to a considerable extent for such tasks as pass point selection, tracing of contours and transfer of planimetric details, which are collectively known as topographic photo-interpretation.[54] Therefore, any advances in the design of photogrammetric instruments to meet automation needs will also require the automation of the whole process of photo-interpretation, although Konecny has pointed out, during a consideration of the possibilities of automating the various phases of photogrammetric plotting, that functions such as selection and interpretation which involve a considerable amount of human decision-making are very difficult to automate.[55]

Despite the difficulty in automating an essentially subjective process of photo-interpretation, some headway has already been made. According to an analysis of the nature of photo-interpretation made by Vink, photo-interpretation has at least two levels of operation: (a) at the lowest level, to detect, select, recognise, and identify the objects imaged on the photograph; and (b) at the highest level, to evaluate the significance (meanings and relationships) of these objects.[56] Obviously, to automate the lowest level of the photo-interpretation process is much easier than to automate the highest level, and indeed one doubts very much whether anybody has ever attempted to automate the highest level at all.

Even the attempt to automate the lowest level of the photo-interpretation process has not been totally successful. Further examination of this level of operation has revealed that the identities of the objects and elements detected in aerial photographs can be established by means of such image characteristics as size, shape, shadow, tone, texture, pattern and position. The process is therefore essentially classificatory because this attempts to sort out all photo images into classes. This classificatory process, however, can be formulated as a problem of the statistical testing of the hypothesis that the two groups of photo images separated by a line in a two-dimensional mathematical space are significantly different. Statistically, this is equivalent to a process

253

of minimising the 'within-group' variances and maximising
the variances between groups.

This appreciation of photo-interpretation as a statistical process of classification has formed the basis
for automation. Such a development is naturally connected with the problems of automatic pattern recognition
which has been a major focus of research by scientists
working in the field of artificial intelligence ever
since 1957. The classical concern of pattern recognition
is to program the computer to recognise printed characters - alphanumerics, punctuation marks, mathematical
symbols, etc.[57] This usually involves the comparison of
a character with an ideal version of each of the possible
characters to see which one it most nearly resembles.
Therefore, the problem of pattern recognition is also the
problem of assigning a name to, or classifying, the many
different characters of a particular class.[58] But of
course the input data involved in pattern recognition
are much less complex than the almost infinite number of
variables of the complex photographic imageries in automatic photo-interpretation. For most classification
purposes, it is necessary to carry out two operations:
(a) to select a set of measurements (features) for the
objects to be classified; and (b) to use these features
either to distinguish between two objects *or*, more
generally, to define the classes.

In automatic photographic classification as a first
level of photo-interpretation, exactly the same two
steps will be performed, but because of the complexity
of the photo variables, an additional step of pre-
processing may also be performed so that one may view
the whole problem in terms of three stages:[59] (1) the
pre-processing stage, (2) the feature extraction stage,
and (3) the decision-making stage.

1 *The Pre-processing Stage*. This involves the
transformation of a given photograph into one or more new
photographs by performing various operations on it. The
purpose is to make the automatic classification easier
by bringing out more sharply certain useful characteristics of the photographed objects. A simple example is
to 'smooth' a photograph in order to suppress distracting
features, or to 'sharpen' it in order to reduce blur.

One common practice is called image-tone enhancement which makes use of some chemical processes to suppress or amplify very small tonal differences in order to enhance the primary image characteristics.[60] This transformed photograph is then put through a special machine for automatic photographic classification or identification.

2 *The Feature Extraction Stage.* The selection of the significant characteristics which best describe the object in quantitative terms is the main task at this stage. There are no hard-and-fast rules governing what and how many features to select, and indeed because of the large number of parameters which may be used for this purpose, the selection has to be left entirely to the intuition and experience of the scientist concerned.[61] However, it has been pointed out that properties should be selected on the basis of their mathematical simplicity or ease of implementation.[62] The soundest basis for property selection, therefore, would be to formulate mathematical descriptions of the classes to which the objects or patterns have to be assigned. Thus, in dealing with shapes and figures, which are quite usual in photo-interpretation, geometrical measurements such as area, height and width are useful properties for pattern classification. As for the number of features to be selected, it is obvious that the larger the number, the more clear-cut will be the classes of objects or patterns defined, but this may not always be possible in practice because of the large storage space required in the computer. The result after this stage of work will be sets of real numbers representing the distinguishing features.

3 *The Decision-Making Stage.* This involves a comparison of the measured features with stored standard examples about the classes. A multivariate statistical procedure will be used for making an automatic assignment of objects into the pre-determined classes. Usually, the discriminant analysis technique is favoured. This makes use of the fact that objects can be represented geometrically as points in an n-dimensional space; the n coordinates of each point are the numerical values of the features selected to represent the object. A classification system can be devised whereby the n-dimensional feature space can be partitioned into *regions*, each of which ideally contains points of only one class.[63] With

255

reference to Fig 6.19, let g_1, g_2, ...,g_m be m possible groups, and let the vector

$$X = \begin{bmatrix} x_1 \\ x_2 \\ \cdot \\ \cdot \\ \cdot \\ x_n \end{bmatrix}$$

be the n features measured on the objects where x_i represents the ith feature measurement, then the discriminant function $D_j(X)$ associated with class g_j, $j=1, 2, ...,m$, is such that if the vector X is in class g_i, the value of $D_i(X)$ must be the largest, ie, for all X belonging to the class g_i,

$$D_i(X) > D_j(X)$$

$$i, j = 1,2,...,m; \ i \neq j$$

Geometrically, in the feature space defined by these n features, the boundary of partition, called the decision boundary, between regions associated with class g_i and g_j, respectively, is expressed by the following equation:

$$D_i(X) - D_j(X) = 0$$

Many different forms of discriminant functions that satisfy the above condition can be selected for $D_i(X)$.

These three different phases of operation have formed the main characteristics of most automatic machines designed for photographic classification or even photo-interpretation purposes. However, it has long been realised that in dealing with photographic images it is necessary to reduce the amount of redundant information by using only a few powerful criteria in the automatic recognition process.[64] There are two major categories of information which have been used in this way: (a) shape and size information, and (b) 'textural' information about the photographs. The 'textural' information in fact gives

Plate 29 A part of the orthophotograph of Kelvingrove area, Glasgow, produced by the GZ-1 orthoprojector with a scan speed of 2.5mm/sec. Note the chopping up of the bridge at (A) and mismatches of the roof of the building at (B) *(Department of Geography, University of Glasgow)*

Plate 30 The Topocart B of Zeiss Jena with the Orthophot *(VEB Carl Zeiss, Jena)*

Plate 31 An example of the orthophotograph and its stereomate of the Renfrew International Test Area produced by the Photogrammetric Research Section, National Research Council, Canada. Note that the stereomodel obtained is metrically correct and direct measurement can be made in height as well as in linear distance. Some cartographic enhancements have been made, especially in the roads, so that the orthophoto is in fact a topographical map (*National Research Council, Canada*)

Plate 32 The Stereocompiler for use with the stereo-orthophotograph developed by the National Research Council of Canada (*National Research Council, Canada*)

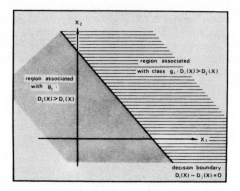

Fig 6.19 *Decision space*

the parameters describing the local distributions of densities and density contrasts in the photographic image.
Automatic recognition of objects with reference to these
criteria is usually carried out by means of the optical
or electronic correlation techniques which generally
involve the matching of the unknown image against some
standard.[65] In particular, the 'textural' information
which in fact is a record of variations in the image
brightness or photographic tone (or optical density) can
be digitised by subjecting the aerial photograph to a
scanner (such as a line-scan cathode ray tube), and as
many as sixty-four separate shades of grey can be
digitised. This method has been seen by Colwell to form
the basis for the development of an automatic photo-
interpretation system, because theoretically the machine
can measure from the digitised scan record such image
characteristics as size, shape, shadow, tone, texture,
pattern and position which are the necessary clues for
photo-identification.[66] But, in practice, this is not
always possible.

In 1961 a machine known as the *Perceptron* was developed
in the United States, employing the method of template
matching which requires pre-processing of the input
photograph.[67] The Perceptron can be 'taught' or 'trained'
to recognise simple graphic forms and has been designed

257

to simulate some portion of the visual perception mechanism of the human brain; but its application to photo-interpretation has been limited by the fact that it cannot separate the photographic images from one another and from the background. But for later machines, even with the use of a micro-densitometer, which is a precise instrument designed for measuring elements of tone and texture of the photograph in continuous scans, it was found to be difficult to develop diagnostic tonal patterns of terrain features because numerous other parameters such as film exposure, processing, and printing can come in to affect the scan patterns, thus disturbing the characteristic density values of such terrain features as vegetation, culture, moisture condition, soil colours, etc.[68] It appeared from research that the use of a multi-channel sensor which can simultaneously obtain up to eighteen channels of imagery ranging from the near ultra-violet to the far infra-red regions offers the greatest potential for automatically delineating various terrain features by means of their normalised spectral response curves.

Modern efforts in developing an automatic system of photo-interpretation have been concentrating on perfecting the photographic (ie optical) as well as electronic systems. The image-tone enhancement technique mentioned earlier is an example of the former. A relatively new approach is the application of optical filtering of the original photograph by coherent light such as laser to produce diffraction patterns at infinity representing the spectrum of the different objects in the photograph. In mathematical terms, a two-dimensional Fourier transform has been done to the original aerial photograph as a pre-processing stage. A simple converging lens is used to transfer the diffraction pattern from infinity to the focal plane where it can easily be either observed on a screen or photographed or even measured quantitatively by means of a photo-electric cell (Fig 6.20). In addition, an image of the original photograph which has been filtered and purified as a result of the transformation brought about in the spectrum can be obtained. Such a method of optical filtering is particularly useful in emphasising alignments and linear features.[69]

For development in the electronic component of the

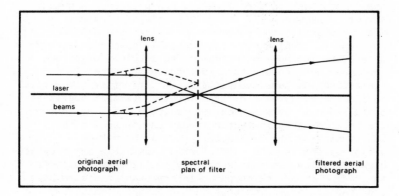

Fig 6.20 *Principle of coherent-light
optical filtering device*

automatic system, more advanced scanners such as the
cathode-ray tube flying spot scanner or camera tubes
such as vidicons or image dissector tubes are employed
together with a more efficient scanning pattern, eg in
the United States, the Itek EC5 Electronic Correlator
which is also the major component of a new optical
stereoscope with automatic image registration.[70] The
stereoscope is very versatile and can take in non-
vertical frame, panoramic and strip photography. The
techniques of image scanning and correlation are used to
measure X- and Y-parallaxes and the crossed diagonal
pattern of scanning (called the Lissajous scanning
pattern) produced by two triangular waveforms of slightly
different frequency is used for more efficient detection
of image displacements.

In the development of an automatic photo-interpretation
system, further improvements of the decision-making stage
of the machine have been made. Thus, the so-called

adaptive network approach is usually preferred to the
fixed network approach. The latter approach assesses
the relative importance of the features by statistical
means, employing known probabilities of occurrence of
each feature as a basis for classification, whilst the
former approach is more flexible and would not assume
any prior knowledge of the relative importance or
distribution of the features comprising each pattern.
An adaptive machine will automatically or progressively
change its internal organisation as a result of being
'trained' by exposure to a succession of correctly
classified and labelled patterns.[71] A further theoretical
refinement of automatic classification is that suggested
by Fu who sets up a sequential decision model so that
the machine takes in feature measurements sequentially
and terminates the sequential process (by making a
decision) only when a sufficient or desirable accuracy of
classification has been achieved.[72] All these attempts
to develop an automatic photo-interpretation system have
been particularly prominent in the United States, but
a fully automatic system has not yet been developed. Di
Pentima has mentioned a system developed by Cornell
Aeronautical Laboratories Inc of Buffalo, New York, which
is really a series of data processing facilities with
provision for pattern recognition and image processing,
and such a system can work in 'real-time' with a general-
purpose computer.[73]

It is clear from all these developments that some progress is being made in those technologies that utilise
or can utilise high-contrast imagery in their research
and can limit their automation efforts to relatively
small populations, eg in the biomedical sciences for the
recognition of cells and in high-energy physics for the
analysis of bubble-chamber photography.[74] But it must
also be realised that, as Di Pentima has rightly observed,
despite considerable effort and work having been expended
towards the development of fully automatic interpretation
systems and components of such systems, little progress
is being made towards the development of automatic interpretation systems that will function with the almost
infinite image variables that are present in even the
simplest of conventional aerial photographs.[75]

Although a fully automatic system of photo-interpretation
is difficult to achieve, yet some degree of automation
is not impossible if a more practical approach is adopted.
Such an approach has been advocated by Steiner and Maurer
in developing a so-called 'quantitative and semi-automatic
system of photo-interpretation of terrain cover type'.[76]
The terrain cover types that they are interested in are
crops. Their approach involves the manual extraction
from aerial photographs of quantitative parameters of
photometric and geometric measurements, such as photographic densities (tones) and height information of crops,
in one or more sample areas where detailed field mapping
has been done as a control. These parameters are then
analysed and classified by means of the multivariate
statistical method of discriminant analysis and the
digital computer. From this a numerical photo-interpretation key is obtained, which can then be applied to
identify crop types in other areas where no field visit
has previously been made. In this way, an objective
classification can be obtained and a reasonable degree of
accuracy of crop classification is usually attainable if
a multitude of parameters are used. Further improvements
in accuracy are possible by extracting more parameters by
means of multi-type photography (such as panchromatic,
infra-red, true- and false-colour) and/or by using photography of different time periods (to take account of seasonal effects in the case of crops).

As illustrations to such a practical semi-automatic approach, Steiner and his collaborators have performed
several experiments employing different types of parameter obtained from aerial photographs for crop classification. One experiment made use of the stereometric
parameters of heights of crops in June and July which
were accurately obtained from large-scale aerial photographs (1:8,000) with a universal stereoplotter, the
Wild A7, and these crop heights were used as the discriminating feature vector (X). A linear discriminant
function was employed for $D_i(X)$ for classification, which
means that the decision boundary is treated as a straight
line. If there are n discriminating variables $X_1, X_2, \ldots$
X_n, then the linear discriminant function Z is given by

261

$$Z = \lambda_1 X_1 + \lambda_2 X_2 + \ldots + \lambda_n X_n$$

where λs are weights for the corresponding discriminating variable Xs.[77] Thus, a computer-generated classification of crops is obtained.

Other parameters which have been used include densito-metric measurements for recording tonal variations which may be made on sets of large-scale simultaneous multi-photography (panchromatic and infra-red black-and-white, and true- and false-colour films) and the temporal variat-ions on such parameters.[78] A combination of all these various parameters in time and space can be used and an improved algorithm of classification and a sequential approach in the inclusion of feature parameters mentioned earlier is possible to produce a more natural classifica-tion. The results have been found to be reasonably accurate. A similar system of crop identification using remotely sensed radar imagery was also reported by Schwarz and Caspall.[79]

Such a semi-automatic approach is not limited to crop classification, but has also found application in urban morphological studies. The classification of buildings in the city into different age groups is possible by making use of such external characteristics of a building as the height, the area, the roof form, the parapet wall, the degree of excrescence, etc. This has been demon-strated by the present author for the city centre of Glasgow, where the buildings have been fairly successful-ly classified automatically into six pre-determined types by age periods as (a) Georgian, (b) Early Victorian, (c) Late Victorian, (d) Edwardian, (e) Modern Inter-War and (f) Modern Post-War Buildings.[80]

It must be noted that the semi-automatic approach is still rather tedious as far as the manual extraction of the feature parameters (which invariably involves stereo-metric or photographic measurements) is concerned. But with advances in automatic instruments such as the digitisers, the quantification of descriptive information for computer input will be an easy matter, and the semi-automatic approach may well answer the need for rapid and more objective identification of objects from aerial photographs, although one should bear in mind that in the

262

use of discriminant analysis, strict objectivity still cannot be achieved since the number and types of features to be selected for each object can only be subjectively determined, so that classification is essentially a subjective procedure employing objective methods.[81] A review of the techniques by Steiner has also clearly demonstrated that different results appear according to the types of methods of discriminant analysis employed.[82] Therefore, one has to exercise great care in using any of these so-called automatic or semi-automatic photo-interpretation systems.

CONCLUSIONS

In this chapter, technical developments towards automation in photogrammetry and photo-interpretation have been reviewed. It can be seen that such developments have come about as a result of the need to extract quantitative and qualitative data rapidly from photographs and non-photographic remotely sensed imageries which were acquired from aircraft or earth-orbiting platforms. It is noteworthy that an increasingly analytical approach is adopted as a result of rapid technological advances in electronics and computer designs, and this gives rise to the only universal system in handling different types of remotely sensed imageries. One important form of development - the orthophotograph, which provides a convenient means of automating the measurement of an individual stereomodel - should be of special importance to geographers. The rapidity with which the orthophotograph can be produced implies that a metrically accurate base containing a wealth of up-to-date planimetric information is readily available to geographers who no longer need to depend solely on the map, the traditional tool, which takes a much longer time to make. Its use in studying rapidly changing phenomena, such as those existing in an urban environment or those observed along the coast, is particularly suitable.[83] From the cartographic point of view, automation has a great economic implication, for the topographic maps which are most urgently required for resources development and planning purposes in the developing countries can now be produced much faster. One of the important implications of this automation trend is perhaps in elevating the non-photographic remote sensing systems to the same level as that of conventional aerial photography and their subsequent integration to form an efficient spatial information-acquiring system for a variety of purposes. On the other hand, the fact that full automation in photogrammetry and photo-interpretation is still not possible at the present stage of technological development is worthy of note because this implies that the discrepancy between the time of acquiring the photographs or imageries (which is very quick) and that of retrieving the data from them (which is very slow) still exists, although the gap has narrowed somewhat with the emergence of orthophotography, image correlation and pattern recognition techniques.

VII TOWARDS INTEGRATION AND UNIFIED GEOGRAPHIC INFORMATION SYSTEMS

Our survey has emphasised two points about aerial photography: (a) it is extremely versatile as a data-collecting and analytical tool in all fields of geography; and (b) it evolves into a unique array of techniques with greatly expanded capabilities and constant refinements. Aerial photography is destined to play an increasingly important role in applied geography. With the modern geographer's concern for more accurate and up-to-date information regarding the terrestrial environment, it is realised that the aerial photographic tool is best utilised through the integrated and systems approaches.

AN INTEGRATED APPROACH: THE EXAMPLE OF LAND EVALUATION

It has already been pointed out in earlier chapters that the metric and descriptive characteristics of aerial photographs are inseparable in actual use; and so are the human and natural features of the terrestrial environment. To the geographer, it is this integrative power of aerial photographs that is most valuable, and hence naturally invites, as de Haas pointed out,[1] an inter-disciplinary approach from physical and social scientists, if the data are to form a basis for making decisions on the planning and implementation of development projects.

An example of the integrated approach is best seen in land evaluation, which is the assessment of the suitability of land for man's use in agriculture, forestry, engineering, hydrology, regional planning, recreation, etc.[2] In essence, the work is largely land classification which is characterised by the delineation of regions, units, etc of uniform characteristics - a familiar method of generalisation in regional geography. According to Mabbutt, there are at least three different approaches to land classification: the genetic, landscape, and

265

parametric methods.[3] The genetic approach makes use of
Herbertson's concept of natural regions delineated on the
basis of climate and structure, but the resultant regions
are too large to be handled and utilised for land assessment purposes. Therefore, the landscape approach which
involves the recognition of the distinctive components
of the landscape is usually preferred. This is the
essence of the 'land system surveys' of the CSIRO in
Australia and is equivalent to the landscape method of
photo-interpretation in the USSR.[4] The basic land unit
of classification is termed the *land facet* by Webster
and Beckett;[5] it should be sufficiently homogeneous over
its extent to be managed uniformly for all but the most
intensive kinds of land use and mappable at scales of
1:10,000 to 1:50,000. However, slightly smaller-scale
aerial photographs may be used for its recognition. It
has been observed that over a large area the same few
land facets recur, usually in a more or less regular
pattern with the same interrelations. This repetition
of a particular set of land facets bestows on the landscape its character. The resultant recurrent landscape
patterns are called *land systems*, which can be mapped
even at much smaller scales than the land facets, in the
range of 1:250,000 to 1,000,000. These maps then provide
the basis for land evaluation. It is useful to draw for
each land system a block diagram to show the characteristic arrangement of the land facets in the landscape
relating both the horizontal and vertical dimensions
together, and the aerial photographs in stereopairs or
triplets should also be properly annotated to show this.
Vertical descriptions are also appended to show the
characteristic landform, materials and hydrology and the
vegetation cover for each distinctive land facet which
forms the overall land system. An example is shown here
of the Chankoko land system in Uganda (Fig 7.1 and Table
7.1).

A similar approach was reported by MacPhail who called
his basic unit of classification a 'photomorphic area',
which is characterised by 'a broad, repetitive pattern
forming the composite image of the fields and fence lines,
the system of drainage, and the tone range of land use,
rock outcrop, soil moisture and vegetation', thus corresponding to the land system of Webster and Beckett.[6]
Photomorphic types can then be identified and mapped.

266

Table 7.1   Chankoko Land System (see Fig 7.1)
(after Webster and Beckett, 1970)

Land facet	Landform	Materials and hydrology	Land cover
1	*Laterite plateau*	shallow humose loam or red clay over laterite; weathered schist, etc beneath; drainage free	*Themeda-Loudetia* savanna
2	*small hills and tors*	granite or gneiss with no soil	nil
3	*quartzite ridges*	quartzite; no soil or very shallow soil	nil, or very sparse savanna
4	*interfluves*	brown or red-brown loam or sandy clay loam over weathered rock; incipient laterite in lower sites; drainage free	dry *Acacia* savanna
5	*valley floors*	sandy loam over dark clay weathered schist, etc at 2-3m depth; swampy; high ground water; locally narrow stream channel	*Themeda-Acacia* savanna

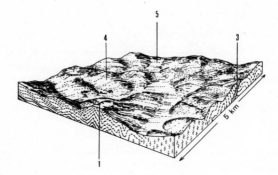

Fig 7.1 *Chankoko land system (see Table 7.1) (after Webster and Beckett, 1970)*

Ground checks of random samples of 2 per cent of photomorphic types identified in Chile confirmed strong correlations with important landscape features and hence the validity of such an approach.

The landscape approach of land classification has made use of landform as the key for reflecting other environmental factors, such as soils, ground and surface water, vegetation, agricultural activities, etc. Verstappen[7] considered this focus of interest on landform to be more efficient than that on vegetation favoured by other workers such as Ray.[8] However, the landscape approach has been criticised as being descriptive and highly subjective, so that the reliability of the classification may be doubted.

The third, and modern, approach is the parametric approach which is aimed at achieving more precise definition of land and avoiding subjectivity through quantification. It relies heavily on the quantitative geomorphology advocated by Horton, Strahler and others, discussed in Chapter III. This means that quantifiable attributes (parameters) must be selected so that they establish land units at component and at pattern level as in the landscape approach. At the component level, form

268

will generally predominate, which includes such parameters as shape, profile and dimension. At the pattern
level, there are two types of parameter: (a) those which
record the dominant or average content of pattern occurrences in respect of any feature, eg average relief or
slope, drainage patterns, and (b) those which express
the arrangement of components, eg measurements of the
overall grain of relief (or dissection) based on slope
changes, the linearity and parallelism exhibited by the
arrangement of peaked and flat-topped upland forms, etc.
For each of the parameters involved, an array of values
for a number of sample points is established, from which
an isopleth map or a best-fit map may be constructed
through interpolation or trend-surface analysis. These
mapped patterns can be combined together to define the
land units - an application of the regionalisation concept viewed as a classification procedure, which can be
automated as shown in Chapter VI, so that a land typology
is established. Such an approach is particularly suited
to the use of aerial photographs and other remotely
sensed non-photographic imageries, and its greatest
potential lies in the possibility of its being automated
by means of the computer.

THE NEED FOR A SYSTEMS APPROACH

The integrated approach implies the need for a *systems
approach* whereby the individual components that form the
complete system have to be taken into consideration to
see how they function efficiently as a whole. Formally,
a system is 'a set of objects together with relationships between the objects and between their attributes'.[9]
As an example, the aerial photograph itself is the product
of a camera system whose components consist of the lens,
shutter, and film. The aerial photographs so obtained
are placed on a stereoplotting machine, which is a photogrammetric system linking the aerial photographs, the
human operator, and the output products (topographic
maps or digital values on tape or cards). Jerie suggested the concept of 'photogrammetric systems engineering' in order to optimise the whole mapping process by
photogrammetric methods.[10] Such a development is deemed
necessary because a very large choice of methods,
equipment and systems is now possible, the scope of
photogrammetry is expanding, and it is realistic to

weight more properly the cost factors in production against purely theoretical considerations. One sees here an economic advantage for the coordination of effort in topographic and non-topographic applications of photogrammetric systems wherever possible. An example is the selection of a suitable scale of photography which can be used to plot topographic maps at the desired scale with the predetermined degree of accuracy as well as for geographic photo-interpretation. Thus, the systems approach is aimed at solving problems characterised by a host of complex phenomena and requirements such as in the evaluation of the human environment. Such a philosophy subsequently leads to the development of information systems.

CHARACTERISTICS OF GEOGRAPHIC INFORMATION SYSTEMS

The information system is developed to meet the need for efficient use of the data which are flowing in from various sources in problem-solving. Thus, it can be defined as 'a group of entities and activities meaningfully connected and satisfactorily bounded which interact with a common purpose or purposes'.[11] For such a system to be operational, it should consist of three sub-systems: the data processing sub-system, the data analysis sub-system and the management sub-system. The data processing subsystem consists of such functions as data acquisition, data input and data storage; the data analysis sub-system carries out retrieval, analysis and information output; and the management sub-system controls the operation of the whole information system. It is also possible to add an information-use sub-system to cater for the needs of different users, thus giving rise to a configuration of the information system as shown in Fig 7.2.[12]

Geographical data are characterised by being location-specific, ie the data have to be fixed in position within an arbitrarily defined coordinate field. Thus, a special information system is required to handle such geographic data. The purpose is 'to make spatially oriented information available in a usable form', which is the map; whilst 'in other and more common situations, the geographic information system is required to reduce several or many map patterns and spatial data sets to words and numbers'.[13] To allow a scientific use of data, there should be a

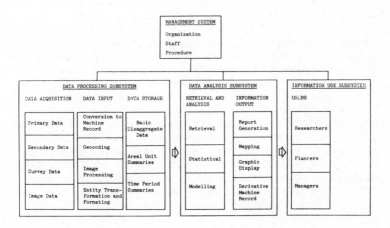

Fig 7.2 *Information system elements*
         (after Tomlinson, 1972)

strong emphasis on 'the interactions between data and
theories, with theory being continually modified and
verified in the light of the data', whilst 'the search
for new data is continually modified by adaptation of
the theory'.[14] The implementation of such a geographic
information system necessitates the use of an electronic
computer with additional graphical output facilities such
as a drum-plotter or a flat-bed coordinatograph. Because
of these special requirements of the geographic information system, data acquired by aerial cameras or other
types of remote sensor are particularly suited for input
into the system. According to Moore and Wellar, the
suitability of these data should be evaluated against
four criteria, which are timeliness, flexibility,
compatibility and reliability.[15]

Timeliness refers to the up-to-dateness of the data
collected, and the major stumbling block is the time-lag
between the acquisition of the imageries and the interpretation to produce meaningful data. It is the

271

characteristic of aerial cameras and other types of remote sensor that temporal changes of phenomena can be detected with ease by comparing the imageries at two points in time. *Flexibility* refers to the power to satisfy the needs of different users with different purposes in mind. The flexibility of the photographs and imageries is especially striking since they are non-selective records of physical and social space; and it is up to the users to extract those data most relevant to their problems from the same imageries. *Compatibility* of all the remotely sensed data for different areas at different times can be controlled through careful cali-bration of the remote sensing systems so that standardised outputs can be achieved. Finally, the criterion of *reliability* refers to the accuracy with which properties of real-world phenomena can be recorded. For the imaging systems, this obviously depends on the sensitivity and capacity of the recording instruments as well as the altitude from which the imageries are obtained. The interpretation is another source of error. Even where 'objective' interpretation algorithm is used in the computer, errors in interpretation can still occur, as shown in Chapter VI. In conclusion, overall evaluation of remotely sensed imageries and photographs has confirmed the usefulness of airborne sensors as data-collecting tools for direct input to the geographic information system.

DEVELOPMENT OF GEOGRAPHIC INFORMATION SYSTEMS BASED ON AERIAL PHOTOGRAPHY: SOME EXAMPLES

Land Use Information System

With the growing interest in the use of aerial photo-graphs and the more general availability of the electronic computer, the possibility of creating some kind of in-formation system has been realised. An early attempt was to apply it to land-use surveys based on aerial photo-graphs. Jeffers reported on such a system which has been designed to provide information on (a) estimates of the proportions of the land surface occupied by each land-use type, together with standard errors of the estimates; (b) estimates of the area occupied by each land-use type, together with standard errors of the estimates; and (c) land-use maps at different scales.[16] The data input was

based on photo-interpretation of sample points superimposed on photographs of the specified area. The data
so obtained were punched on cards for feeding into the
computer. They were then subjected to a fairly intensive
and systematic check by the computer under a special
programme control before being analysed. Coordinates of
each of the sample points could be worked out, and
graphical outputs from the system were possible by means
of the computer-controlled digital incremental plotter.
The flow diagram for this land-use information system
is shown in Fig 7.3.

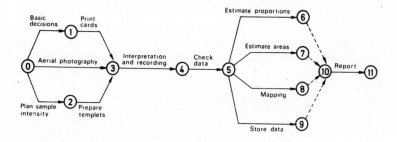

Fig 7.3 *Flow chart of different stages of
land-use survey (after Jeffers, 1967)*

*Terrain Information System: The Digital Terrain Model
Concept*

Recent developments in photogrammetric instrumentation
have made possible the development of a terrain information system based on the Digital Terrain Model (DTM)

concept. This requires the use of a stereoplotter with
automatic digitising facilities. The stereomodel is
reconstructed with the aid of the stereoplotter and the
height data at points are recorded on tape by the
digitiser at the same time as the contours are being
drawn. These height data are then stored in the computer
and represent the terrain model in digital form. But
it will not be possible to digitise the stereomodel
continuously, since the digitiser is of only limited
resolution in recording. The DTM can be regarded as a
statistical representation of the continuous surface of
the ground by a large number of selected points with
known X-, Y-, Z-coordinates in an arbitrary coordinate
field, according to the conceptualisation of Miller and
Laflamme.[17] The great advantage of the DTM lies in the
flexibility with which the terrain data can be called
up in a variety of forms for analysis, and this allows
the evaluation of an unlimited number of independent
solutions to each type of problem. Thus, the computer
can be programmed to carry out height interpolation based
on the sample of digitised points on the surface of the
stereomodel, or to construct terrain profiles along
different directions. It is small wonder that this concept has received much attention from engineers. To
geographers, the DTM concept has even more generality,
since the Z-values do not necessarily refer to terrain
heights. They can be socio-economic variables, for
example, the degree of poverty or the probability of
migration. Provided the storage capacity of the computer
is large enough, additional data can be extracted from
aerial photographs through the photogrammetric plotting
machine and stored with the DTM. An actual application
of such a concept is seen in the highway information
system developed by Colner.[18] This system extracts and
stores data on the name, the geographical location, the
number and nature of the nodes (ie junctions), and the
width and the street furniture (eg parking meters, fire
hydrants, etc) of the road as well as some limited
building-use information found along the road. Such a
system is in fact an integrated photogrammetric data
bank which can collect, store, update and retrieve data
to help the highway engineers cope with problems in highway planning, design, construction and maintenance. Since
the system is connected to automated line plotters,
graphical output is possible. The design of this

photogrammetric data bank is shown in Fig 7.4.

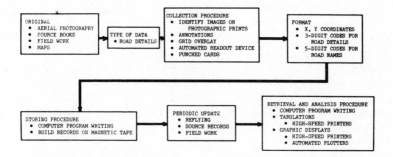

Fig 7.4 *Photogrammetric data bank
concept for highway engineering (after
Colner, 1966)*

Urban Information Systems

Much attention has been drawn towards the development of
urban information systems by town planners and geographers. This is not surprising in view of the increasing
complexity of the urban system. There is an urgent demand
for up-to-date data for use in the day-to-day administra-
tion as well as the long-range planning of the city, and
an urban information system is the most efficient means of
coordinating data from different sources, regularly updating them, and quickly retrieving, displaying and organ-
ising them for preliminary analysis. For such an urban
information system to be developed, there are even more
stringent requirements for integration and flexibility
than other information systems. Dueker and Horton

have illustrated these problems with reference to an urban-change detection system which they have designed (Fig 7.5).[19] Clearly, this information system comprises the four sub-systems of data management, processing,

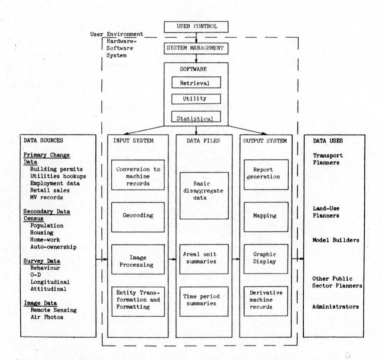

Fig 7.5 *Urban planning information system elements with an emphasis on urban-change detection (after Dueker and Horton, 1972)*

analysis and use mentioned before. One should note that highly varied data from ground surveys, documents as well as imageries, are input to this system. There are three system elements which need to be considered: (a) the method of spatial location, (b) the frequency of observation, and (c) the kind of phenomena to be observed.

The method of spatial location or geocoding refers to the means of specifying location in a machine-readable form. This is usually direct digitising of observations to X- and Y-coordinates, especially when photographs, maps or other graphics are the data source. Another method involves assigning areal-unit codes to observations, eg when the data come from ground surveys. The third method is unique addressing by which street addresses are translated either to geographic coordinates or to areal-unit codes by referring to a Geographic Base File in the computer, ie a directory of address ranges assigned to give an areal unit for comparison to the input address. In order to relate data from different sources to a common geographic base, a system of *small* areal units based on geographic coordinates is found to be most suitable.

The second system element concerns the frequency of updating the data base which obviously depends on the use of the data and how frequently the specific phenomena observed have changed status. Where constant updating and maintenance are required, airborne sensors should be used for the collection of these data at regular time intervals.

The third system element, ie the kind of phenomena to be observed, concerns the social, economic and physical phenomena which should all be measured and integrated into the information system files. All three types of system elements together with the data sources should be integrated into the information system in order to provide current data about social, economic and physical variables by small areas for planning and management. Dueker and Horton have stressed the usefulness of conventional aerial photography as data inputs to such an information system. In particular, the versatility of colour infra-red photography was noted. They have also demonstrated the suitability of such data inputs for five specific items: (a) transport networks, (b) land use, (c) population, (d) dwelling-unit counts, and (e) economic activity, all of which involved the technique of photo-interpretation explained in Chapter III.

An example of an actually implemented urban information system based on remote sensing inputs has been examined by Wellar.[20] This is the Wichita Falls, Texas, Municipal Information System. The data inputs were aerial photographs of different scales, which were employed for traffic flow and accident studies, determination of the most feasible alignments for street construction, and the preparation of street plans. Smaller-scale aerial photography was used to show the growth patterns of the city, to construct base maps for land-use data banks and to prepare photo-mosaics for engineering planning purposes. It was concluded that data inputs and analyses will continue to be based on aerial photography and photogrammetry for some time, which will involve a mix of manual and automated methods of data generation, processing, dissemination and application.

In this connection, one should take note of the potential of orthophotography examined in Chapter VI. It should form a part of the output sub-system in the urban information system since orthophotography permits an up-to-date map-like data base of the city to be produced directly from aerial photographs.[21] This development has enhanced the importance of the photogrammetric plotting machine as a self-contained information system when coupled to a computer. The stereo-orthophotography system introduced by Collins and Blachut further simplifies the use of orthophotographs by geographers while still maintaining in their mind a geometrically correct three-dimensional stereomodel (as opposed to the affine deformed stereomodel of unrectified aerial photographs).[22]

The Geographic Information System of the Rural Development Branch, Canada

The concept of the geographic information system has been implemented in the Department of Forestry and Rural Development in Canada since 1963 to provide data for guiding the development of its land, water and human resources. This can meet the need for assembling social, economic and land data for an integrated analysis of the problems of rural development.[23] The computerised system allows map and related data to be stored in a form suitable for rapid measurement and comparison, which would not have been possible with manual handling, owing to the very great amount of data. The system can accept and store all types of location-specific data, ie any data that can be related to an area, line or point on a map. It comprises two parts: the data bank and the set of procedures and methods for moving data into the bank, and for carrying out the manipulations, measurements, and comparison of the data. The information required can be output both in numerical form by the line-printer and in map form by a graphic plotter. The procedures of data preparation for the system are shown in Fig 7.6.

The whole system can also accept inputs from aerial photographs or other types of imageries. The input data from aerial photographs can be rectified by an analytical approach with the computer. On the other hand, ortho-photographs should prove to be even more suitable as a data source for direct input to the system, especially for the purpose of updating.

CONCLUSIONS

In this concluding chapter, an attempt has been made to illustrate the two most important characteristics of aerial photography, namely, its integrating power and its suitability as data inputs to geographic information systems. The development of information systems is indicative of the increasing emphasis by geographers on computerisation and automation in decision-making, and in data generation, processing, dissemination and application. A more scientific geography is possible only when data are readily available for testing the truth of hypotheses and for building models of real-world phenomena.

The further expansion of aerial photography in recent years has coincided with a period when geographers have been paying increasing attention to the well-being of our living environment. One the one hand, there is the need to search for more natural resources for the developed and developing countries alike, whilst, on the other hand, mankind is becoming more concerned with maintaining the balance of our ecosystem. Space

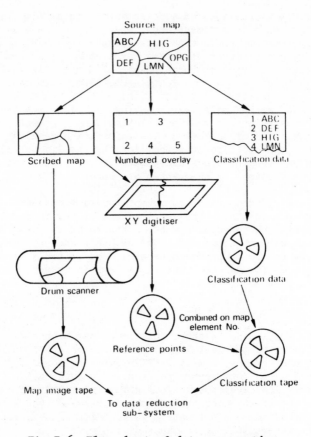

Fig 7.6 *Flow chart of data-preparation procedures of the Canadian Geographic Information System (after Tomlinson, 1968)*

exploration by the more affluent countries has brought new platforms for cameras so that truly holistic views of our Earth are for the first time available to geographers. Space photography thus provides a vantage-point which supplements the bird's-eye-view vantage-point of conventional aerial photography. What is more, we now have at our disposal a real-time surveillance system which puts our Earth under constant watch, thus allowing us to monitor day-to-day changes that have occurred in our terrestrial environment.

Amidst all the changes in the philosophical conceptions of geography, one is struck by the flexibility of aerial photography in meeting the very varied needs of geographers. It has emerged even more prominently than before as a data-collecting and analytical tool. In view of its versatility, one is certain that there will be increased activity in the use of aerial photography by geographers in different research fields in the years to come.

NOTES

CHAPTER I

1 Eliel, L.T. 'One Hundred Years of Photogrammetry', *Photogrammetric Engineering*, 25 (1959), 356-63. Also see Mekel, J.F.M., *Geology from the Air*, (Delft, 1969)

2 Lee, W.T. *The Face of the Earth as seen from the Air*. Special Publication No 4, American Geographical Society (1922)

3 Trager, H. 'Precision Lenses and Shutter', *Photogrammetric Engineering*, 22 (1956), 912-20

4 Eliel. *op cit*, 362

5 Colwell, R.N. *Manual of Photographic Interpretation*, American Society of Photogrammetry (1960), 735-92

6 Grigg, D. 'Regions, Models and Classes' in Haggett, P. and R.J. Chorley (eds), *Models in Geography* (1967), 461-509

7 McPhail, D.C. 'Photomorphic Mapping in Chile', *Photogrammetric Engineering*, 37 (1971), 1139-48

8 Peplies, R.W. 'Regional Analysis and Remote Sensing' in Estes, J.E., and L.W. Senger (eds), *An Introductory Reader in Remote Sensing*, International Geographical Union, Montreal Meeting (1972), III. 1-18

9 Raasveldt, H.C. 'The Stereomodel, How it is Formed and Deformed', *Photogrammetric Engineering*, 22 (1956), 708-26

10 Bedwell, C.H. 'The Eye, Vision and Photogrammetry', *Photogrammetric Record*, 7 (1971), 135-56

11 Lo, C.P. 'Modern Use of Aerial Photographs in Geographical Research', *Area*, 3 (1971), 164-9

CHAPTER II

1 Thompson, M.M. (ed). *Manual of Photogrammetry*, Vol I, American Society of Photogrammetry (1966), 1

2 *Ibid*. Vol II, 875-9

3 War Office, *Manual of Map Reading, Air Photo Reading and Field Sketching*, Pt II: *Air Photo Reading*, HMSO (1958), 15-18

4 Ziemann, H. 'Image Geometry - Factors Contributing to its Change', Invited Paper, Commission I, 12th International Congress of Photogrammetry in Ottawa, Canada (1972)

5 Langford, M.J. *Advanced Photography - A Grammar of Techniques* (1972), 161

6 Howard, J.A. *Aerial Photo-Ecology* (1970), 39

7 Mullins, L. 'Some Important Characteristics of Photographic Materials for Air Photographs', *Photogrammetric Record*, 5 (1966) 240-70

8 *Ibid*. 253; see also Langford, *op cit*, 58

9 See Welch, R. 'Modulation Transfer Functions', *Photogrammetric Engineering*, 37 (1971), 247-59, and Rikets Allmänna Kartverk (Geographical Survey Office of Sweden), 'Photographic Modulation Transfer Functions', Report by the Commission I Working Group of the International Society for Photogrammetry to the 12th International Congress for Photogrammetry in Ottawa, Canada, 1972. See also Mullins. *op cit*, 253-9. A detailed treatment of the subject can be found in Brock, G.C. *Image Evaluation for Aerial Photography* (1970)

10   Rikets Allmänna Kartverk. *op cit*, 10-12

11   Adelstein, P.Z.   'Dimensional Stability of Estar
     Base Films', *Photogrammetric Engineering*, 38 (1972)
     55-64

12   Eden, J.A.   'The Unsharp Mask Technique of Printing
     Aerial Negatives', *Photogrammetric Record*, 1, No 5
     (1955), 5-26; Craig, D.R.   'The Logetronic Strip
     Printer Model SP 10/70', *Photogrammetric
     Engineering*, 24 (1958), 780-6

13   Corten, F.L.   *Image Quality in Air Photography*,
     ITC Publications, Series A/B No 5, Vol I
     (Delft, 1960)

14   Ziemann.   *op cit*, 37

15   Sherishen, A.I.   *Aerial Photography*, Israel Program
     for Scientific Translations (Jerusalem, 1961),
     200-5

16   Heimes, F.J.   'In-Flight Calibration of a Survey
     Aircraft System', Paper presented to Commission I,
     12th International Congress for Photogrammetry,
     Ottawa, Canada, 1972

17   Swanson, L.W.   'Aerial Photography and Photogram-
     metry in the Coast and Geodetic Survey',
     Photogrammetric Engineering, 30 (1964), 699-726

18   Schermerhorn, W.   'The Use of Airborne Instruments
     in the Determination of Absolute Orientation
     Elements', *Photogrammetric Record*, 5 (1966), 271-88

19   Kennedy, D.   'Airborne Photographic Equipment',
     Photogrammetric Engineering, 31 (1965), 971-7

20   Eden, J.A.   'The Airborne Profile Recorder',
     Photogrammetric Record, 2 (1957), 263-78

21   Trott, T.   'Development of Aerial Camera Stabilisa-
     tion', *Photogrammetric Engineering*, 23 (1957), 122-
     30.   See also his 'Verticality in Photogrammetry',
     Photogrammetric Engineering, 24 (1958), 63-9

22   Thompson. *op cit*, 156-8

23   *Ibid.* 145-56. See also Hovey, S.T., 'Panoramic Possibilities and Problems', *Photogrammetric Engineering*, 31 (1965), 727-34

24   Colwell, R.N. *Manual of Aerial Photographic Interpretation*, American Society of Photogrammetry (1960), 138

25   Moffitt, F.H. *Photogrammetry* (1967), 79

26   Crone, D.R. *Elementary Photogrammetry* (1963), 10-11; Kilford, W.K. *Elementary Air Survey*, (1973). See also Thompson, *op cit*, 803

27   National Physical Laboratory. *Modern Computing Methods, Notes on Applied Science*, No 12, HMSO (1965), 5-12

28   Thompson, E.H. 'Height from Parallax Measurements', *Photogrammetric Record*, 1 (1954), 38-49

29   Kilford, W.K. *Elementary Air Survey* (1973), 228-37

30   Methley, B.D.F. 'Heights from Parallax Bar and Computer', *Photogrammetric Record*, 6 (1970), 459-65. The full program written in Algol is described in his *A Computer Program for Parallax Bar Heighting*, Department of Geography, University of Glasgow (1971)

31   Methley. 'Heights from Parallax Bar ...', 462-3

32   Newton, I. 'Flight Commissioning and Ground Control', in Goodier, R. (ed), *The Application of Aerial Photography to the Work of the Nature Conservancy*, The Nature Conservancy (Edinburgh, 1971), 110-19

33   Visser, J. *Radial Triangulation*, ITC Textbook, Vol III.2 (Delft, 1968)

34   Kilford. *op cit*, 98-119

35 Moffitt. *op cit*, 144-72

36 Kubik, K., and D. Tait. *Photogrammetric Triangulation in Space with Analogue Instruments*, ITC Lecture Notes (1967)

37· Ambrose, W.R., and R.T. Stone. 'New Cartographic Techniques with the Zoom Transfer Scope', *Papers from the 1971 ASP-ACSM Fall Convention*, American Society of Photogrammetry (1971), 110-24. See also Yzerman, H. 'The Z.T.S., a Unique Instrument for the Transfer of Photographic Detail Onto a Reference Map', *Technical Bulletin*, Bausch and Lomb Inc (1972)

38 Moffitt. *op cit*, 315-21

39 Wolf, P.R. *Elements of Photogrammetry* (1974), 296-300

40 Jones, A.D. 'Photogrammetric Techniques in Geography' in Bowen, E.G., H. Carter and J.A. Taylor (eds), *Geography at Aberystwyth*, University of Wales, Cardiff (1968), 143-9

41 Thompson, E.H. 'A New Photogrammetric Plotter: the CP-1', *Photogrammetric Record*, 7 (1971), 157-81. See also his 'The CP-1 Plotter: Setting Procedure and Results', *Ibid*, 7 (1972), 323-33

42 Thompson, M.M. (ed). *Manual of Photogrammetry*, Vol 1, American Society of Photogrammetry (1966), 461

CHAPTER III

1 Miller, V.C. 'Aerial Photographs and Landforms (Photogeomorphology)', in *Aerial Surveys and Integrated Studies*, (UNESCO 1968), 41-69. See also his *Photogeology*, (1961)

2 Miller. 'Aerial Photographs', 41

3 Verstappen, H. Th. *Fundamentals of Photogeology/Geomorphology*, ITC Textbook, Vol VII (Delft 1963)

4   Ray, R.G. *Aerial Photographs in Geologic Inter-*
    pretation and Mapping,(United States Government
    Printing Office 1960)

5   Ray, R.G. and W.A. Fischer. 'Quantitative Photo-
    graphy - A Geologic Research Tool', *Photogrammetric*
    Engineering, 26 (1960), 143-50

6   Stellingwerf, D.A., and J.M. Remeyn. *Measurements*
    and Estimations on Aerial Photographs for Forest
    Purposes, ITC Textbook (Delft, 1969), 100

7   Blanchet, P.H. 'Development of Fracture Analysis
    as Exploration Methods', *Bulletin of the American*
    Association of Petroleum Geologists, 41 (1957),
    1748-59

8   Lattman, L.H., and R.P. Nickelsen. 'Photographic
    Fracture-Trace Mapping in Appalachian Plateau',
    Bulletin of American Association of Petroleum
    Geologists, 42 (1958), 2238-45

9   See recent works by (1) Boyer, R.E., and J.E.
    McQueen, 'Comparison of Mapped Rock Fractures and
    Airphoto Linear Features', *Photogrammetric*
    Engineering, 30 (1964), 630-5; (2) Norman, J.W.
    'Linear Geological Features as an Aid to Photo-
    geological Research', *Photogrammetria,* 25 (1969/70),
    177-87

10 Stephens, E.A. 'Structural Analysis from Air
    Photographs in Areas of Regionally Metamorphosed
    Rocks', *Photogrammetria,* 25 (1969/70), 5-26

11 Verstappen, H. Th. *Fundamentals of Photogeology/*
    Geomorphology (I), ITC Textbook, Vol VIII, 1
    (Delft, 1963)

12 Mekel, J.F., J.F. Savage and H.C. Zorn. *Slope*
    Measurements and Estimates from Aerial Photographs.
    ITC Publication, B/26, (Delft, 1967)

13 *Ibid.* 8-14

14 Ray. *op cit,* 63

15 Hackman, R.J. 'The Stereo Slope Comparator - An Instrument for Measuring Angles of Slope in Stereoscopic Models', *Photogrammetric Engineering*, 22 (1956), 387-91

16 Stellingwerf and Remeyn. *op cit*, 87

17 *Ibid.* 89

18 Verstappen. *op cit*, 42

19 Desjardins, L. 'The Measurement of Formational Thickness by Photogeology', *Photogrammetric Engineering*, 17 (1951), 821-31

20 Ray and Fischer. *op cit*, 147-8

21 Horton, R.E. 'Erosional Development of Streams and their Drainage Basins: Hydrophysical Approach to Quantitative Geomorphology', *Bulletin of Geological Society of America*, 56 (1945), 275-370

22 Strahler, A.N. 'Statistical Analysis in Geomorphic Research', *Journal of Geology*, 12 (1954), 1-25. See also his *Dimensional Analysis in Geomorphology*, (Columbia University, 1957)

23 For a full discussion, see Haggett, P., and R.J. Chorley, *Network Analysis in Geography* (1969), 10-16

24 Shaw, S.H. 'The Value of Aerial Photographs in the Analysis of Drainage Pattern', *Photogrammetric Record*, 1 (1953), 4-20

25 Bunik, J.A., and A.K. Turner, 'Remote Sensing Applications to the Quantitative Analysis of Drainage Networks', *Papers from the 1971 ASP-ACSM Fall Convention*, American Society of Photogrammetry (1971), 166-78

26 Finsterwalder, R. 'Photogrammetry and Glacier Research with special reference to Glacier Retreat in the Eastern Alps', *Journal of Glaciology*, 2 (1954), 306-15

27 Case, J.B. 'Mapping of Glaciers in Alaska'
    Photogrammetric Engineering, 24 (1958), 815-21

28 Konecny, G. 'Glacial Surveys in Western Canada',
    Photogrammetric Engineering, 30 (1964), 64-82.
    See also his 'Applications of Photogrammetry to
    Surveys of Glaciers in Canada and Alaska',
    Canadian Journal of Earth Sciences, 3 (1964),
    783-98

29 Petrie, G., and R.J. Price. 'Photogrammetric
    Measurements of the Ice Wastage and Morphological
    Changes near the Casement Glacier, Alaska',
    Canadian Journal of Earth Sciences, 3 (1966),
    827-40

30 Welch, R., and P.J. Howarth. 'Photogrammetric
    Measurements of Glacial Landforms', *Photogrammetric
    Record*, 6 (1968), 75-96

31 El-Ashry, M.T., and H.R. Wanless. 'Shoreline
    Features and their Changes', *Photogrammetric
    Engineering*, 33 (1967), 184-9

32 Alexander, C.S. 'A Method of Descriptive Shore
    Classification and Mapping as applied to the
    North East Coast of Tanganyika', *Annals*, Association
    of American Geographers, 56 (1966), 128-40

33 Jones, A.D. 'Aspects of Comparative Air Photo
    Interpretation in the Dyfi Estuary', *Photogrammetric
    Record*, 6 (1969), 291-305

34 Hubbard, J.C.E. 'The Use of Aerial Photography in
    the Survey of Coastal Features' in Goodier, R. (ed),
    *The Application of Aerial Photography to the Work
    of the Nature Conservancy*, Nature Conservancy
    (Edinburgh, 1971), 36-42

35 Walker, F. *Geography from the Air* (1964), 63-72

36 Sonu, C.J. 'Study of Shore Processes with aid of
    Aerial Photogrammetry', *Photogrammetric Engineering*,
    30 (1964), 932-41

37   Cameron, H.L. 'Water Current and Movement Measurement with Time-Lapse Air Photography: An Evaluation', *Photogrammetric Engineering*, 28 (1962), 158-63

38   Forrester, W.D. 'Plotting of Water Current Patterns by Photogrammetry', *Photogrammetric Engineering*, 26 (1960), 726-36

39   Keller, M. 'Tidal Current Surveys by Photogrammetric Methods', *Photogrammetric Engineering*, 29 (1963), 824-32

40   Komarov, V.B. 'Aerial Photography in the Investigation of Natural Resources in the USSR', in *Aerial Surveys and Integrated Studies*, UNESCO (1968), 143-85

41   Finkel, H.J. 'The Movement of Barchan Dunes Measured by Aerial Photogrammetry', *Photogrammetric Engineering*, 27 (1961), 439-44

42   Bagnold, R.A. *The Physics of Blown Sand and Desert Dunes* (1954)

43   Clos-Arceduc, A. 'Emploi des photographies aeriennes pour l'étude des dunes Sahariennes allongées dans une direction voisine de celle du vent', *Photogrammetria*, 25 (1969/70), 189-99

44   Orville, H.D. 'Cumulus Cloud Photogrammetry', *Photogrammetric Engineering*, 27 (1961), 787-91

45   Boge, W.E. 'Upper Atmospheric Wind Determination from Stereo Photography of Rocket Vapour Trails', *Photogrammetric Engineering*, 29 (1963), 1059-68

46   Pietschner, J. 'Determination of the Wind Vector in the Lower Atmosphere by Stereophotogrammetric Surveys of Smoke Marks' *Photogrammetric Record*, 7 (1971), 223-31

47   Tansley, A.G. 'The Use and Abuse of Vegetational Concepts and Terms', *Ecology*, 16 (1935), 284-307

48   Howard, J.A. *Aerial Photo-Ecology* (1970), 259

49 Greig-Smith, P. *Quantitative Ecology* (1964)

50 Husch, B. *Forest Mensuration and Statistics* (1963)

51 Parry, J.T., and C.M. Gold. 'A Solar-Altitude Nomogram', *Photogrammetric Engineering*, 38 (1972), 891-9

52 Ericsson, H. 'Concerning Accuracy in Measuring Tree and Stand Heights', *International Archives of Photogrammetry*, 13 (1961), Pt 6, *et seq*, 138

53 Howard. *op cit*, 251

54 *Ibid.*

55 Loetsch, F., and E. Haller. 'The Adjustment of Area Computations from Sampling Devices on Aerial Photographs', *Photogrammetric Engineering*, 28 (1962), 789-810

56 Stellingwerf and Remeyn. *op cit*, Chapters 3 and 4

57 Spurr, S.H. *Photogrammetry and Photo-Interpretation* (1960), 152-65

58 *Ibid.* 159-64

59 Wilson, R.C. 'The Relief Displacement Factor in Forest Area Estimates by Dot Templets on Aerial Photographs', *Photogrammetric Engineering*, 15 (1949), 225-326

60 Stellingwerf and Remeyn. *op cit*, Chapter 6

61 Rogers, E.J. 'Application of Aerial Photographs and Regression Technique for Surveying Caspian Forests of Iran', *Photogrammetric Engineering*, 26 (1961), 441-3

62 Howard. *op cit*, 258-71

63 Conzen, M.R.G. 'Historical Townscapes in Britain: A Problem in Applied Geography', in House, J.W. (ed),

Northern Geographical Essays in Honour of G.H.J. Daysh (1966), 56-78

64   Whitehand, J.W.R. 'Building Types as a Basis for Settlement Classification' in Whittow, J.B., and P.D. Wood (eds), *Essays in Geography for Austin Miller*, University of Reading (1965), 291-305

65   Johns, E. *British Townscapes* (1965)

66   Lo, C.P. 'Aerial Photographic Analysis of the Urban Environment: A Study of the Three-Dimensional Aspects of Land Use in the City Centres of Glasgow and Hong Kong', Unpublished PhD Thesis, University of Glasgow (1971), 122-6

67   Collins, W.G., and A.H.A. El-Beik. 'Population Census with the aid of Aerial Photographs: An Experiment in the City of Leeds', *Photogrammetric Record*, 7 (1971), 16-26

68   Hsu, S.Y. 'Population Estimation', *Photogrammetric Engineering*, 37 (1971), 449-54

69   Eyre, L.A., B. Adolphus and M. Amiel. 'Census Analysis and Population Studies', *Photogrammetric Engineering*, 36 (1970), 460-6

70   Pryor, W.T. 'Highway Engineering Applications of Photogrammetry', *Photogrammetric Engineering*, 20 (1954), 523-31

71   Stoch, L., *et al.* 'Urban Traffic Surveys from Aerial Photographs', *The South African Journal of Photogrammetry*, 3 (1970), 275-85

72   Wohl, M., and S.M. Sickle 'Continuous Strip Photography - An Approach to Traffic Studies', *Photogrammetric Engineering*, 25 (1959), 397-403

73   Taylor, J.I. 'Photogrammetric Determination of Traffic Flow Parameters', *International Archives of Photogrammetry*, Vol 17, Part 10 (1969), Commission V, 5.32

74 St Joseph, J.K. 'Air Photography and Archaeology', in J.K. St Joseph (ed), *The Uses of Air Photography* (1966), 113-25

75 Fairhurst, H., and G. Petrie. 'Scottish Clachans II: Lix and Rosal', *Scottish Geographical Magazine*, 80 (1964), 150-63

76 See a review by Atkinson, K.B. 'Some Recent Developments in Non-Topographic Photogrammetry', *Photogrammetric Record*, 6 (1969), 357-78

77 Moffitt, F.H. *Photogrammetry* (1967); see Chapter 14, 448-525, for a detailed discussion of the method

78 Thompson, E.H. 'Photogrammetry in the Restoration of Castle Howard', *Photogrammetric Record*, 4 (1962), 94-119

79 The progress of the Commission since 1963 was reported on by McDowall, R.W. 'Uses of Photogrammetry in the Study of Buildings', *Photogrammetric Record*, 7 (1972), 390-404

80 See a short report on this in *Wild Reporter*, No 5 (September 1972), p 10

81 Hardegen, L. 'The Application of Photogrammetry to the Conservation of Monuments', *Schweizerische Technische Zeitchrift*, 66 (1969), 721-31. See also Brucklacher, W. 'An Equipment System for Architectural Photogrammetry', Paper presented to the 12th International Congress for Photogrammetry, Ottawa, Canada, 1972

82 See Löschner, F., D. Berling and H. Foramitti. 'Architektur-photogrammetrie Denkmalpflege - Kulturgüterschutz', Paper presented to Commission V of the 12th International Congress for Photogrammetry, Ottawa, Canada, 1972. See also Aston, B., and C. Sena. 'Considerations on the Photogrammetric Survey of the Castle of Vigevano', Paper presented to Commission V of the 12th International Congress for Photogrammetry, Ottawa,

Canada, 1972

83 Cheffins, O.W., and J.E.M. Rushton. 'Edinburgh
    Castle Rock: A Survey of the North Face by
    Terrestrial Photogrammetry', *Photogrammetric Record*,
    6 (1970), 417-33. See also Reid, I.A. 'Survey and
    Mapping of the Rock Face of American Niagara Falls',
    Department of Environment, Ottawa, Canada (1972)

CHAPTER IV

1 Tait, D.A. 'Photo-Interpretation and Topographic
   Mapping', *Photogrammetric Record*, 6 (1970), 466-79

2 De Haas, W.G.L. 'Aerial Survey and Social Space',
   *Transactions of the Second International
   Symposium on Photo-Interpretation*, Paris, 1966,
   V.41-V.44

3 Colwell, R.N. (ed). *Manual of Photographic
   Interpretation*, American Society of Photogrammetry
   (1960), 1

4 Kudritskii, D.M., I.V. Popov and E.A. Romanova,
   Hydrographic Interpretation of Aerial Photographs,
   Israel Program for Scientific Translations
   (Jerusalem, 1966), 84

5 Vink, A.P.A. *Some Thoughts on Photo-Interpretation*,
   ITC Publication, Series B/25 (Delft, 1964), 7-13

6 *Ibid.* 13

7 Shearer, J.W. *Topography*, ITC Lecture Notes, 4

8 *Ibid.* 2; see also Rabben, E.L., in Colwell, *op cit*,
   99-105; and Ray, R.G., *Aerial Photographs in Geologic Interpretation and Mapping*, United States
   Government Printing Office (1960), 6-13

9 Colwell. *op cit*, 112-13

10 De Lancie, R., *et al.* 'Quantitative Evaluation of
    Photo-Interpretation Keys', *Photogrammetric
    Engineering*, 23 (1957), 858-64

11 Sayn-Wittgenstein, L. 'Recognition of Tree Species
    on Air Photographs by Crown Characteristics',
    Photogrammetric Engineering, 27 (1961), 792-809

12 Chisnell, T.C., and G.E. Cole. '"Industrial
    Components" - A Photo Interpretation Key on
    Industry', *Photogrammetric Engineering*, 24 (1958),
    590-602

13 Powers, W.E. 'A Key for the Photo-Identification
    of Glacial Landforms', *Photogrammetric Engineering*,
    17 (1951), 776-9

14 Waldo, C.A., and R.P. Ireland. 'Construction of
    Landform Keys', *Photogrammetric Engineering*, 21
    (1955), 603-6

15 Heath, G.R. 'A Comparison of Two Basic Theories
    of Land Classification and their Adaptability to
    Regional Photo Interpretation Key Techniques',
    Photogrammetric Engineering, 22 (1956), 144-68

16 Lewis, G.K. 'The Concept of Analogous Area Photo
    Interpretation Key', *Photogrammetric Engineering*,
    23 (1957), 874-8

17 Bigelow, G.F. 'Photographic Interpretation Keys -
    A Reappraisal', *Photogrammetric Engineering*, 29
    (1963), 1042-51

18 McCue, G.A., and J. Green. 'Pisagh Crater Terrain
    Analysis', *Photogrammetric Engineering*, 31 (1965),
    810-21

19 Verstappen, H. Th. 'Some Observations on Karst
    Development in the Malay Archipelago', *Journal of
    Tropical Geography*, 14 (1960), 1-10. See also
    his 'Karst Morphology of the Star Mountains (Central
    New Guinea) and its Relation to Lithology and
    Climate', *Zeitschrift für Geomorphologie*, 8 (1964),
    40-9; Marker, M. 'Some Problems of a Karst Area in
    the Eastern Transvaal, South Africa', *Transactions*,
    Institute of British Geographers, 50 (1970), 73-
    85

20   Price, R.J.   'The Changing Proglacial Environment
      of the Casement Glacier, Glacier Bay, Alaska',
      Transactions, Institute of British Geographers,
      36 (1965), 107-16. See also his 'Moraines,
      Sandar, Kames and Eskers near Breidamerkurjökull,
      Iceland', *Ibid.* 46 (1969), 17-37

21   Dishaw, H.E.   'Massive Landslides', *Photogrammetric
      Engineering*, 33 (1967), 603-8. See also Cazabat, C.
      'Les cartes de localisation probable des avalanches',
      Paper presented to the 12th International Congress
      for Photogrammetry, Ottawa, Canada, 1972, which
      described the mapping of probable avalanche
      concentration areas in the Alps and Pyrenees by the
      Institut Geographique National in France

22   Shaw, S.H.   'The Value of Air Photographs in the
      Analysis of Drainage Patterns', *Photogrammetric
      Record*, 1 (1953), 4-20. See also Tator, B.A.
      'Drainage Anomalies in Coastal Plains Regions',
      Photogrammetric Engineering, 20 (1954), 412-17;
      Nossin, J.J., 'The Geomorphic History of the
      Northern Pahang Delta', *Journal of Tropical
      Geography*, 20 (1965), 54-64

23   El-Ashry, M.T., and H.R. Wanless.   'Birth and
      Early Growth of a Tidal Delta', *Journal of Geology*,
      73 (1965), 404-6. See also their 'Shoreline
      Features and their Changes', *Photogrammetric
      Engineering*, 33 (1967), 184-9; Stafford, D.B., and
      J. Langfelder, 'An Aerial Photographic Survey of
      Coastal Erosion', *Papers* from 36th Annual Meeting
      of the American Society of Photogrammetry, 1970,
      533-8. Finally, a useful recent volume is:
      Shepard, F.P., and H.R. Wanless, *Our Changing
      Coastline* (1971)

24   Ritchie, W.   'The Machair of South Uist', *Scottish
      Geographical Magazine*, 83 (1967), 161-73; Kesik, A.
      'Photo Interpretation of Loess Landscape in Poland',
      Photogrammetria, 26 (1970), 183-200

25   Davis, C.K., and J.T. Neal.   'Descriptions and
      Airphoto Characteristics of Desert Land forms',
      Photogrammetric Engineering, 29 (1963), 621-31

26 Sharpe, C.F.S. *Landslides and Related Phenomena*, Columbia University Press (1938). See also Thornbury, W.D., *Principles of Geomorphology* (1954), 44-6

27 Lueder, D.R. *Aerial Photographic Interpretation* (1959)

28 Davis and Neal. *op cit*, 622

29 Verstappen, H. Th. *Aerial Photographs in Geology and Geomorphology*, ITC Textbook of Photo-Interpretation (Delft, 1963), 9

30 Zernitz, E.R. 'Drainage Patterns and their Significance', *Journal of Geology*, 40 (1932), 498-521

31 Kudritskii, D.M., I.V. Popov and E.A. Romanova. *Hydrographic Interpretation of Aerial Photographs*, Israel Program for Scientific Translations (Jerusalem, 1966), 99-121

32 Colwell. *op cit*; Verstappen, *op cit*; Lueder, *op cit*; Ray, R.G. *Aerial Photographs in Geologic Interpretation and Mapping*, United States Government Printing Office (1960); Allum, J.A.E. *Photogeology and Regional Mapping* (1966)

33 Nunnally, N.R. 'Interpreting Land Use from Remote Sensor Imagery', in Estes, J.E., and L.W. Senger (eds), *An Introductory Reader in Remote Sensing* (Montreal, 1972), VI-1 - VI-21

34 Clawson, M. 'Recent Efforts to Improve Land Use Information', *Journal of American Statistical Association*, 61 (1966), 647-57

35 Stamp, L.D. *Applied Geography* (1960), 37-50

36 Coleman, A., and K.R.A. Maggs. *Land Use Survey Handbook*, Isle of Thanet Geographical Association (1965)

37 Anderson, J.R. 'Land-Use Classification Schemes',

Photogrammetric Engineering, 37 (1971), 379-87

38 Clawson, *op cit*, 652-5

39 Haggett, P. *Locational Analysis in Human Geography* (1965), 191-200

40 Berry, B.J.L., and A.M. Baker, 'Geographic Sampling' in Berry B.J.L., and D. Marble (eds), *Spatial Analysis: A Reader in Statistical Geography* (1968), 91-100

41 Goodman, M.J. 'A Technique for the Identification of Farm Crops on Aerial Photographs', *Photogrammetric Engineering*, 25 (1959), 131-7

42 Ciolkosz, A. 'Analysis of Crop Structure on Aerial Photographs', Paper presented to Commission VII of the 12th International Congress for Photogrammetry, Ottawa, Canada, 1972

43 Goodman. *op cit*, 134-5

44 Goodman, M.S. 'Criteria for the Identification of Types of Farming on Aerial Photographs', *Photogrammetric Engineering*, 30 (1964), 984-90

45 Goodier, R., and B.H. Grimes. 'The Interpretation and Mapping of Vegetation and Other Ground Surface Features from Air Photographs of Mountainous Areas in North Wales', *Photogrammetric Record*, 6 (1970), 553-66. See also Ward, S.D., *et al*, 'The Use of Aerial Photography for Vegetation Mapping and Vegetation Interpretation' in Goodier, R. (ed), *The Application of Aerial Photography to the Work of Nature Conservancy*, Nature Conservancy (Edinburgh, 1971), 52-65

46 Kudritskii, *et al*. *op cit*, 129

47 *Ibid*. 130-2

48 Goodier and Grimes. *op cit*, 560

49 Sayn-Wittgenstein. *op cit*, 797-809

50   Komarov, V.B.   'Aerial Photography in the Investi-
     gation of Natural Resources in the USSR',
     Aerial Surveys and Integrated Studies, UNESCO
     (1968), 157-68

51   Dill, H.W., Jr.   'Use of the Comparison Method in
     Agricultural Air Photo Interpretation',
     Photogrammetric Engineering, 25 (1959), 44-9

52   Avery, G.   'Measuring Land Use Changes on USDA
     Photographs', *Photogrammetric Engineering*, 31 (1965),
     620-4

53   Witenstein, M.W.   'A Report on Application of Aerial
     Photography to Urban Land-Use Inventory, Analysis,
     and Planning', *Photogrammetric Engineering*, 22
     (1956), 656-64

54   Wray, J.R.   'Photo Interpretation in Urban Area
     Analysis', in Colwell, R.N. (ed), *op cit*, 677-80

55   Collins, W.G., and A.H.A. El-Beik.   'The Acquisition
     of Urban Land Use Information from Aerial Photo-
     graphs of the City of Leeds', *Photogrammetria*,
     27 (1971), 71-92

56   Turpin, R.D.   'Evaluation of Photogrammetry and
     Photographic Interpretation for Use in
     Transportation Planning', *Photogrammetric
     Engineering*, 30 (1964), 124-30

57   Richmond, I.   'Towns and Monumental Buildings', in
     St Joseph, J.K.S. (ed), *The Use of Aerial Photo-
     graphy, Nature and Man in a New Perspective* (1966),
     138-49

58   Lynch, K.   *The Image of the City* (1960)

59   Dale, P.F.   'Children's Reactions to Maps and
     Aerial Photography', *Area*, 3 (1971), 170-7

60   Howlett, B.E.   'Determining Urban Growth and
     Changes from Aerial Photograph Comparison',
     Highway Research Record, 19 (1963), 1-16

61 Wagner, R.R. 'Using Air Photos to Measure Changes
in Land Use around Highway Interchanges',
Photogrammetric Engineering, 29 (1963), 645-9

62 Falkner, E. 'Land Use Changes in Parkway School
District', *Photogrammetric Engineering*, 34 (1968),
52-7

63 Richter, D.M. 'Sequential Urban Change', *Photogrammetric Engineering*, 35 (1969), 764-70

64 Simonett, D.S. 'Present and Future Needs of Remote
Sensing in Geography', *Proceedings of the Fourth
Symposium on Remote Sensing of Environment*,
University of Michigan (1966), 37-45

65 Quoted by Simakova, M.S. *Soil Mapping by Colour
Aerial Photography*, Academy of Sciences of the
USSR, English translation (Jerusalem, 1964), 22

66 Frost, R.E. 'Photo Interpretation of Soils', in
Colwell (ed), *op cit*, 343-402

67 Simakova. *op cit*, 60

68 *Ibid.* 22

69 Kudritskii. *op cit*, 128

70 Simakova. *op cit*, 11

71 Evans, R. 'Air Photographs for Soil Survey in
Lowland England: Soil Patterns', *Photogrammetric
Record*, 7 (1972), 302-22

72 Webster, R., and I.F.T. Wong. 'A Numerical
Procedure for Testing Soil Boundaries Interpreted
from Air Photographs', *Photogrammetria*, 24 (1969),
59-72

73 Branch, M.C. *Aerial Photography in Urban Planning
and Research*, Harvard University Press (1948)

74 Green, N.E. 'Aerial Photographic Interpretation and
the Social Structure of the City', *Photogrammetric*

Engineering, 23 (1957), 89-96. See also his 'Aerial Photographic Interpretation and the Human Geography of the City', *Professional Geographer*, 9 (1957), 2-5

75 See, for example, McGee, T.G. *The Southeast Asian City* (1967); and *The Urbanization Process in the Third World: Explorations in Search of a Theory* (1971)

76 Wellar, B.S. *Generation of Housing Quality Data from Multiband Aerial Photographs*, Research Report, Department of Geography, Northwestern University (1967)

77 Mumbower, L.E., and J. Donoghue. 'Urban Poverty Study', *Photogrammetric Engineering*, 33 (1967), 610-18

78 Metivier, E.D., and R.M. McCoy. 'Mapping Urban Poverty Housing from Aerial Photographs', *Proceedings of the Seventh International Symposium on Remote Sensing of Environment*, University of Michigan (1971), 1563-9

CHAPTER V

1 Colwell, R.N. 'Spectrometric Considerations Involved in Making Rural Land Use Studies with Aerial Photography', *Photogrammetria*, 20 (1965), 15-33

2 Jones, A.D. 'Aspects of Comparative Air Photo-Interpretation in the Dyfi Estuary', *Photogrammetric Record*, 6 (1969), 291-305

3 Sonu, C.J. 'Study of Shore Processes with Aid of Aerial Photography', *Photogrammetric Engineering*, 30 (1964), 932-41

4 Steiner, D. 'Technical Aspects of Air Photo Interpretation in the Soviet Union', *Photogrammetric Engineering*, 29 (1963), 988-98. See also Stellingwerf, D.A. *Practical Applications of Aerial Photographs in Forestry and Other Vegetation*

Studies, ITC Publications, Series B, no 36 (Delft, 1966), 11

5   Eastman Kodak Co. *Applied Infra-red Photography*, Kodak Technical Publication M-200 (1970)

6   It was formerly believed that chlorophyll was the infra-red reflective component of the leaf as suggested by such papers as Tarkington, R.G., and A.L. Sorem, 'Color and False Color Films for Aerial Photography', *Photogrammetric Engineering*, 29 (1963), 88-95; and Cooper, C.F., and F.M. Smith, 'Color Aerial Photography: Toy or Tool?', *Journal of Forestry*, 64 (1966), 373-8. But recent studies have confirmed that the reflection in the near infra-red spectral region is more dependent on the *structure* of leaf tissues, whilst chlorophyll and other leaf pigments are more responsible for the reflection of leaves in the visible portions of the spectrum (ie panchromatic photography). See a discussion by Benson, M.L., *et al*, 'The Truth About False Colour Film', *Photogrammetric Record*, 6 (1970), 446-51. A more up-to-date discussion of this problem can be found in Wiegend, C.L., H.W. Gausman and W.A. Allen. 'Physiological Factors and Optical Parameters as Bases of Vegetation Discrimination and Stress Analysis', in *Operational Remote Sensing*, American Soceity of Photogrammetry (1972), 82-100

7   Colwell. *op cit*, 23. See also Andreucci, E., 'The Behaviour of Vegetation in Infra-red Photography', *Ferrania*, 12 (1964), 13-19

8   Eastman Kodak Co. *op cit*, 14-16

9   Stellingwerf. *op cit*, 10

10 Colwell. *op cit*, 25

11 *Ibid.* 25

12 Parry, J.T., and H. Turner. 'Infra-red Photos for Drainage Analysis', *Photogrammetric Engineering*, 37 (1971), 1031-8

13 Jones, B.G. 'Low-Water Photography in Cobscook
    Bay, Maine', *Photogrammetric Engineering*, 23 (1957),
    338-42

14 Theurer, C. 'Color and Infra-red Experimental
    Photography for Coastal Mapping', *Photogrammetric
    Engineering*, 25 (1959), 565-9

15 Jones, A.D., 'Aspects of Comparative Air Photo-
    Interpretation in the Dyfi Estuary', *Photogrammetric
    Record*, 6 (1969), 297

16 Swanson, L.W. 'Photogrammetric Surveys for
    Nautical Charting: The Use of Color and Infra-red
    Photography', *Photogrammetric Engineering*, 26
    (1960), 137-41. See also Brewer, R.K., and
    A.K. Heywood, 'Coastal Boundary Mapping',
    *Proceedings of the 38th Annual Meeting of the
    American Society of Photogrammetry* (1972), 182-
    91

17 Lattman, L.H. "Geologic Interpretation of Airborne
    Infra-red Drift from Infra-red Imagery', *Photo-
    grammetric Engineering*, 22 (1963), 83-7

18 Winkler, E.M. 'The Interpretation of Glacial
    Drift from Infra-red Films', *Photogrammetric
    Engineering*, 26 (1960), 773-4

19 Welch, R. 'A Comparison of Aerial Films in the
    Study of the Breidamerkur Glacier Area, Iceland',
    Photogrammetric Record, 5 (1966), 289-306

20 Maruyasu, T., and M. Nishio. 'On the Study and
    Application of Infra-red Aerial Photography',
    Report of the Institute of Industrial Science,
    University of Tokyo, Vol 10 (1960), 1-16

21 Welch, R. 'The Application of Aerial Photography
    to the Study of a Glacial Area: Breidamerkur,
    Iceland', unpublished PhD thesis, University of
    Glasgow (1967)

22 Lo, C.P. 'Aerial Photographic Analysis of the
    Urban Environment: A Study of the Three-

Dimensional Aspects of Land Use in the City Centres
of Glasgow and Hong Kong', unpublished PhD thesis,
University of Glasgow, (1971)

23 Strandberg, C.H. 'Photoarchaeology', *Photogrammetric Engineering*, 33 (1967), 1152-7

24 Eastman Kodak Co. *Kodak Infra-red Films*, Kodak
Publication No N-17 (1971)

25 Strandberg, C.H. 'The Sensation of Color' in
Smith, J.T., Jr (ed), *Manual of Color Aerial
Photography*, American Society of Photogrammetry
(1968), 3-11

26 Mott, P.G. 'Some Aspects of Colour Aerial Photo-
graphy in Practice and its Applications',
Photogrammetric Record, 5 (1966), 221-37. See also
Woodrow, H.C. 'The Use of Colour Photography for
Large Scale Mapping', *Photogrammetric Record*, 5
(1968), 433-77

27 Meier, H.K. 'Color-Correct Aerial Photography?'
Translation of paper published in *Bildmessung und
Luftbildwesen*, No 5 (1967), 206-14

28 Duddek, M. 'Wild Mapping Cameras for Color' in
Smith, J.T., Jr (ed), *op cit*, 154-82

29 Bormann, G.E. 'The New Wild RC 10 Universal Film
Camera', *Papers from the 35th Annual Meeting of
the American Society of Photogrammetry* (1969), 1-9

30 See (1) Harris, W.D., B.F. Lampton and M.J. Umbach.
'Metric Measurement of Color Photography', in
Smith, J.T., Jr (ed), *op cit*, 324-33; (2) Anson,
A., and J.R. Quick, 'Collection of Metric Color
Photography of the Pheonix, Arizona Test Area',
Ibid, 334-41; (3) Schallock, G.W., 'Metric Tests
of ASP-Phoenix Color Photography', *Ibid*, 342-64

31 English, J.S. 'Vertical Colour – Air Survey's
New Weapon', *Hunting Group Review*, No 3 (1965), 1-8.
See also Tarling, L.W. 'Some Observations and
Recommendations for the Future of Aerial Colour

304

Photography', *Photogrammetric Record*, 6 (1970), 480-3

32 Welch, R. 'Film Transparencies vs Paper Prints',
    Photogrammetric Engineering, 36 (1968), 490-501

33 Anson, A. 'The Application of Color and Multi-
    spectral Techniques to the Collection of Military
    Geographic Information'. Paper presented to the
    12th International Congress for Photogrammetry,
    Ottawa, Canada, 1972

34 See Smith, J.T., Jr *op cit*, Chapter X: 'Photographic
    Interpretation from Color Aerial Photographs',
    381-440

35 Becking, R.W. 'Forestry Applications of Aerial
    Colour Photography', *Photogrammetric Engineering*,
    25 (1959), 559-65

36 Steiner, D., *et al.* quoted by Colwell, R.N., in
    Smith, J.T., Jr (ed), *op cit*, 382-3

37 Wert, S.L., and B. Roettgering. 'Aerial Color
    Photography and Probability Sampling Techniques
    Evaluate Douglas-Fir Beetle Epidemic in California',
    Papers from the 34th Annual Meeting, American
    Society of Photogrammetry, 1968, 348-64

38 Johnson, P.L. 'Radioactive Contamination to
    Vegetation', *Photogrammetric Engineering*, 31 (1965),
    984-90

39 Gerbermann, A.H., H.W. Gausman and C.L. Wiegand.
    'Color and Color-IR Films for Soil Identification',
    Photogrammetric Engineering, 37 (1971), 359-64

40 Steiner, *et al.* *op cit*

41 Jones, A.D. 'Film Emulsions', in Goodier, R. (ed),
    *The Application of Aerial Photography to the Work
    of the Nature Conservancy* (1971), 126

42 Simakova, M.S. *Soil Mapping by Color Aerial
    Photography*, Israel Program for Scientific

Translations (Jerusalem, 1964), 56-8

43 Verstappen, H. Th. *Aerial Photographs in Geology
and Geomorphology*, ITC Textbook, Vol VII, Pt II,
Chapter VII.1 (Delft, 1963), 16-20

44 Kent, B.H. 'Experiments in the Use of Colour
Aerial Photographs for Geologic Study',
Photogrammetric Engineering, 23 (1957), 865-8

45 Trollinger, W.V. 'The Role of Color Photography in
Natural Resources Exploration', Paper presented
to the 12th International Congress for Photogrammetry, Ottawa, Canada, 1972

46 Fischer, W.A. 'Color Aerial Photography in
Geologic Investigations', *Photogrammetric
Engineering*, 28 (1962), 133-9

47 Swanson. *op cit*

48 Theurer. *op cit*

49 Keene, D.F., and W.G. Pearcy. 'High-Altitude
Photographs of the Oregon Coast', *Photogrammetric
Engineering*, 39 (1973), 163-76

50 Strandberg, C.H. 'Aerial Photographic Interpretation Techniques for Water Quality Analysis',
Photogrammetric Engineering, 32 (1966), 234-48

51 Dolan, R., and L. Vincent. 'Coastal Processes',
Photogrammetric Engineering, 39 (1973), 255-60

52 Specht, M.R., D. Needler and N.L. Fritz. 'New
Color Film for Water-Photography Penetration',
Photogrammetric Engineering, 39 (1973), 359-69

53 Welch, R. 'A Comparison of Aerial Films in the
Study of the Breidamerkur Glacier Area, Iceland',
Photogrammetric Record, 5 (1966), 289-306

54 Jones, A.D. 'Aspects of Comparative Air Photo-
Interpretation in the Dyfi Estuary', *Photogrammetric
Record*, 6 (1969), 291-305

55 See, for example, the following papers: (1) Stallard,
    A.H., and R.R. Biege Jr. 'An Evaluation of Color
    Aerial Photography in Some Aspects of Highway
    Engineering', *Highway Research Record*, 109 (1965),
    27-38; (2) Strandberg, C.H. 'Photoarchaeology',
    Photogrammetric Engineering, 33 (1967), 1152-7

56 Cooke, R.U., and D.R. Harris. 'Remote Sensing of
    the Terrestrial Environment - Principles and
    Progress', *Transactions*, Institute of British
    Geographers, 50 (1970), 12

57 Mumbower, L.E., and J. Donoghue. 'Urban Poverty
    Study', *Photogrammetric Engineering*, 33 (1967),
    610-18

58 Manji, A.S. *Uses of Conventional Aerial Photo-
    graphy in Urban Areas: Review and Bibliography*,
    Research Report No 41, Dept. of Geography,
    Northwestern University (1968)

59 Kracht, J.B., and W.A. Howard. *Applications of
    Remote Sensing, Aerial Photography and Instrumented
    Imagery Interpretation to Urban Studies*, Council
    of Planning Librarians Exchange Bibliography, No
    166 (1970)

60 Stellingwerf, D.A. 'The Usefulness of Kodak
    Ektachrome Infra-red Aero Film for Forestry
    Purposes', *International Archives of Photogrammetry*,
    Vol 17, Pt 5 (1969), Ref No 7, 34

61 Benson, W.L., and W.G. Sims. 'False-Colour Film
    Fails In Practice', *Journal of Forestry*, 65 (1967),
    904. See also their 'The Truth About False Colour
    Film: An Australian View', *Photogrammetric Record*,
    6 (1970), 446-8

62 Colwell, R.N. 'Uses of Aerial Photography in
    Agriculture', in Smith, J.T., Jr (ed) *op cit*,
    382-3

63 Hildebrandt, G., and H. Kenneweg. 'The Truth
    About False Colour Film: A German View',
    Photogrammetric Record, 6 (1970), 448-50

64 Wiegand, C.L., H.W. Gausman and W.A. Allen.
    'Physiological Factors and Optical Parameters as
    Bases of Vegetation Discrimination and Stress
    Analysis', in *Operational Remote Sensing*, American
    Society of Photogrammetry (1972), 82-100

65 Arnberg, W., and L. Wastenson, 'Use of Aerial
    Photographs for Early Detection of Bark Beetle
    Infestation of Spruce', *Ambio*, 3 (1973), 77-83

66 Murtha, P.A. 'Classification of Forest Damage
    from Air Photos', Paper presented to Commission
    VII, 12th Congress of International Society for
    Photogrammetry, Ottawa, Canada, 1972

67 Rhode, W.G., and C.E. Olson Jr. 'Detecting Tree
    Moisture Stress with Infra-red Sensors', *Papers
    from the 35th Annual Meeting*, American Society of
    Photogrammetry (1969), 185-90

68 Anderson, R.R., and F.J. Wobber. 'Wetlands Mapping
    in New Jersey', *Photogrammetric Engineering*, 39
    (1973), 353-8

69 Turner, R.M. 'Measurement of Spatial and Temporal
    Changes in Vegetation from Color-IR Film', *Papers
    from the 1971 ASP-ACSM Fall Convention*, American
    Society of Photogrammetry (1971), 426-41

70 Cooke and Harris. *op cit*, 13

71 Samol, J.D. *Rural Land Use Analysis via Ektachrome
    Infra-red Photo Interpretation*, Technical Report
    No 4, East Tennessee State University Remote
    Sensing Institute (no date)

72 Jones. *op cit*, 303

73 Pressman, A.E. 'Geologic Comparison of Ektachrome
    and Infra-red Ektachrome Photography', in Smith J.T.,
    Jr (ed) *op cit*, 397

74 Gerbermann, *et al*. *op cit*, 363

75 Jones, A.D. 'Film Emulsions', in Goodier, R. (ed),

The Application of Aerial Photography to the Work of the Nature Conservancy (1971), 127-8

76 Strandberg, C.H., in Smith, J.T. Jr (ed), *op cit*, 421

77 Hannah, J.W. *A Feasibility Study for the Application of Remote Sensors to Selected Urban and Regional Land Use Planning Studies*, Technical Report No 11, East Tennessee State University Remote Sensing Institute (1968)

78 Marble, D.F., and F.E. Horton. 'Extraction of Urban Data from High and Low Resolution Images', *Proceedings of the 6th International Symposium on Remote Sensing of Environment* (1969), 807-18

79 Lindgren, D.T. 'Dwelling Unit Estimation with Color-IR Photos', *Photogrammetric Engineering*, 37 (1971), 373-7

80 Green, N.E. 'Aerial Photographic Interpretation and the Social Structure of the City', *Photogrammetric Engineering*, 23 (1957), 89-96. Also see Binsell, R., *Dwelling Unit Estimation from Aerial Photography*, Department of Geography, Northwestern University (1967)

81 Colwell, R.N. 'Spectrometric Considerations Involved in Making Rural Land Use Studies with Aerial Photography', *Photogrammetria*, 20 (1965), 15-33

82 Colwell, R.N. 'Aerial and Space Photographs as Aids to Land Evaluation', in Stewart, G.A. (ed), *Land Evaluation*, Macmillan of Australia (1968), 325

83 Itek Optical Division. *Multispectral Photographic Capabilities.* Itek 72-0122-1, Itek Corporation (1972), 7-14

84 Pestrong, R. 'Multiband Photos for a Tidal Marsh', *Photogrammetric Engineering*, 35 (1969), 435-70

85   Yost, E., and S. Wenderoth. 'Multispectral Color
      Aerial Photography', *Photogrammetric Engineering*,
      33 (1967), 1020-33. See also their 'Agricultural
      and Oceanographic Applications on Multispectral
      Color Photography', *Proceedings of the 6th
      International Symposium on Remote Sensing of
      Environment*, University of Michigan (1969),
      145-73

86   Helgeson, G.A. 'Water Depth and Distance
      Penetration', *Photogrammetric Engineering*,
      36 (1970), 164-72

87   Anson. *op cit*

88   Tanguay, M.G., R.M. Hoffer and R.D. Miles.
      'Multispectral Imagery and Automatic Classification
      of Spectral Response for Detailed Engineering
      Soils Mapping', *Proceedings of the 6th International
      Symposium on Remote Sensing of Environment* (1969),
      33-63

89   Orr, D.G., and J.R. Quick. 'Construction Materials
      in Delta Areas', *Photogrammetric Engineering*, 37
      (1971), 337-51

90   Marshall, R.E., N. Thomson, F. Thomson and F.
      Kriegler. 'Use of Multispectral Recognition
      Techniques for Conducting Rapid, Wide-Area
      Wheat Surveys', *Proceedings of the 6th International
      Symposium on Remote Sensing of Environment*,
      University of Michigan (1969), 3-20

91   Meyer, M.P., and H.C. Chiang. 'Multiband
      Reconnaissance of Simulated Insect Defoliation
      in Corn Fields', *Proceedings of the 7th
      International Symposium on Remote Sensing of
      Environment*, University of Michigan (1971),
      1231-4

92   Lauer, D.T., A.S. Benson and C.M. Hay. 'Multiband
      Photography - Forestry and Agricultural
      Applications', *Papers from the 1971 ASP-ACSM Fall
      Convention*, American Society of Photogrammetry,
      531-53

93 Tueller, P.T., and G. Lorain. 'Environmental Analysis of the Lake Tahoe Basin from Small Scale Multispectral Imagery', *Proceedings of the 7th International Symposium on Remote Sensing of Environment*, University of Michigan (1971), 453-67

94 Vinogradov, B.V., A.A. Grigoryev and K. Ya. Konratyev. 'Some Results of Multispectral Aerophysical Measurements of the Earth's Surface at the Usturt Plateau', *Proceedings of the 7th International Symposium on Remote Sensing of Environment*, University of Michigan (1971), 231-6. See also Wetland Mapping Team, 'Aerial Multiband Wetlands Mapping', *Photogrammetric Engineering*, 38 (1972), 1188-9

95 Harris, *et al. op cit*, 324-33

96 Woodrow, H.C. 'The Use of Colour Photography for Large Scale Mapping', *Photogrammetric Record*, 5 (1967), 433-60

97 Schallock. *op cit*, 342-64

98 Reeves, F.B. 'Investigation of the Relative Merits of Black-and-White Versus Color Aerial Photography for a Large Scale Commercial Mapping Project', in *New Horizon in Color Aerial Photography* (American Society of Photogrammetry and Society of Photographic Scientists and Engineers, 1969), 165-74

99 Pope, R.B. 'Effect of Photo Scale on the Accuracy of Forest Measurements', *Photogrammetric Engineering*, 23 (1957), 869-73

100 Lo. *op cit*, 485

CHAPTER VI

1 Haralick, R.M. 'Adaptive Pattern Recognition of Agriculture in Western Kansas by using a Predictive Model in Construction of Similarity Sets', *Proceedings of the Fifth Symposium on Remote Sensing of Environment*, University of Michigan

(1968), 343-56

2 Barrett, R.P. 'Man-Machine Task Sharing in
   Advanced Photo Interpretation Systems',
   *Transactions of the Symposium on Photo Interpreta-
   tion* (Delft, 1962), 52-9

3 Gurk, H.M. 'The Need for More Information and Less
   Data', *Symposium Proceedings on Management and
   Utilization of Remote Sensing Data*, American
   Society of Photogrammetry (1973), 513-27

4 Petrie, G. 'Developments of the Orthophoto and
   Automatic Map Compilation Systems', *Proceedings
   of Symposium on Modern Survey Techniques*,
   Trinity College, Dublin (1968), 58-65

5 Petrie, G. 'Photogrammetric Digitising: Input for
   Data Processing', *Proceedings of the ISP Commission
   IV Symposium* (Delft, 1971), 59-93. See also a
   more up-to-date review in his 'Digitising of
   Photogrammetric Instruments for Cartographic
   Applications', *Photogrammetria*, 28 (1972), 145-71

6 Berger, F., D. Gut and P.B. Stewardson. 'The Wild
   EK8 Coordinate Recording System, and its Potentials',
   Paper presented to Commission IV, 12th International
   Congress for Photogrammetry, Ottawa, Canada, 1972

7 Petrie, *op cit* (1971), 73-5

8 Bailey, K.V., *et al.* 'Bendix Datagrid Digitiser',
   Paper presented at ASP - ACSM Convention, 1969

9 Petrie, G. 'Numerically Controlled Methods of
   Automatic Plotting and Draughting', *Cartographic
   Journal*, 3 (1966), 60-73

10 Anonymous, *Proceedings of the Symposium on Map and
   Chart Digitising*, US Geological Survey Computer
   Contribution No 5 (1970), 5-7

11 Gardiner-Hill, R.C. 'Automated Cartography in the
   Ordnance Survey', Paper presented in the Common-
   wealth Survey Officers Conference, 1971

12 *Ibid.* See also Williams, E.P.J., 'Digitized Ordnance Survey Maps', *Geographical Magazine*, 44 (1972), 780-1

13 Thompson, M.M. (ed). *Manual of Photogrammetry*, Vol 1, American Society of Photogrammetry (1966), 1-11

14 Cimerman, V.J., and Z. Tomasegovic, *Atlas of Photogrammetric Instruments* (1970), 181-9

15 Helava, U. 'A Fast Automatic Plotter', *Photogrammetric Engineering*, 32 (1966), 58-66

16 Cimerman and Tomasegovic, *op cit*, 192-4

17 Makarovic, B. 'Recent Instrument Development in Europe and Asia', Paper presented to Commission II, 12th International Congress for Photogrammetry, Ottawa, Canada, 1972, 20-2

18 Inghilleri, G. 'A New Analytical Plotter: The Digital Stereocartograph (D.S.)', Paper presented to Commission II, 12th International Congress for Photogrammetry, Ottawa, Canada, 1972

19 Dorrer, E., and B. Kurz. 'Plotter Interfaced with a Calculator', *Photogrammetric Engineering*, 39 (1973), 1065-76

20 Van Zuylen, L. 'Production of Photomaps', *Cartographic Journal*, 6 (1969), 92-102

21 See Lacmann, O. 'Entzerrungsgerat fur nicht ebenes Gelande', *Bildmessung und Luftbildwesen*, No 6 (1931), 11-12; and Ferber, R., 'Obtention photographique de la projection orthogonale d'un object', *Bull. Photogrammetrie*, No 3 (1933), 45-53

22 Bean, R.K. 'The Orthophotoscope and its Development', *The Canadian Surveyor*, 22 (1968), 38-46. See also Löscher, W. 'Some Aspects of Orthophoto Technology', *Photogrammetric Record*, 5 (1967), 419-32

23 Pölzleitner, F., 'The Wild PPO-8 Orthophoto

Equipment', Paper presented to Commission II, 12th
International Congress for Photogrammetry, Ottawa,
Canada, 1972

24   Ferri, W.S.   'The Galileo-Santoni Orthophoto-
Simplex and Some Accuracy Tests', Paper presented
to Commission II, 12th International Congress
for Photogrammetry, Ottawa, Canada, 1972

25   Danko, J.O., Jr.   'The Kelsh K-320 Orthoscan - A
New Concept in Orthophotography', *Papers from the
Orthophoto Work Shop II*, American Society of
Photogrammetry (1973), 16-34

26   Helava, U.V.   'On Different Methods of
Orthophotography', *The Canadian Surveyor*, 22 (1968),
5-20

27   Meier, H.K.   'The Gigas-Zeiss Orthoprojector Design
Features and Practical Results', *The Canadian
Surveyor*, 22 (1968), 57-64; see also Meier, H.K.
'Orthoprojection Systems and their Practical
Potential', Paper presented at the Fourth National
Survey Conference, Durban, 1970

28   Weibrecht, O.   'Simultaneous Differential
Rectification and Orography using the "Stereotrigomat" System', *Jena Surveying News*, No 19
(1968), 27-33

29   Szangolies, K.   'Application of Topocart, Orthophot
and Orograph', *Jena Review*, No 2 (1969), 115-20

30   Van Zuylen.   *op cit*, 94-5

31   Ahrend, M., *Large Photogrammetric Instruments*,
Carl Zeiss, Oberkochen (1969), 20.   See also
Visser, J.   'Production of Ortho-photographs',
ITC Paper (Delft, 1968)

32   For tests on planimetric and heighting accuracy of
orthophotographs and drop-line charts, see (1)
Visser, J.   'The Use of and Experience with the
Zeiss Orthoprojector GZ-1 at the ITC', *The Canadian
Surveyor*, 22 (1968), 177-93; (2) Forstner, R.

'Experience with the GZ-1 Orthophoto Projector',
ibid, 106-17; and (3) Johansson, O. 'Orthophoto
Maps as a Basis for the Economic Map of Sweden on
the Scale 1:10,000', *ibid*, 149-58

33 Ferri. *op cit*

34 Meier. *op cit* (1970), 3-4

35 *Ibid.* 7-10

36 Hobbie, D., 'Gigas-Zeiss Orthoprojector Equipment
News', Paper presented to Commission II, 12th
International Congress for Photogrammetry, Ottawa,
Canada, 1972

37 Hughes, T.A., A.R. Shope and F.S. Baxter. 'USGS
Automatic Orthophoto System', *Photogrammetric
Engineering*, 37 (1971), 1055-62

38 Visser, J., *et al.* 'Performance and Applications
of Orthophotomaps', Paper presented to the 12th
International Congress for Photogrammetry,
Ottawa, 1972

39 Drobyshev, F.V. 'Differential Rectification of
Aerial Photos in the USSR', Paper presented to
the 11th International Congress for Photogrammetry,
Lausanne, 1968

40 Makarovic, B. 'Semi-Automatic Mapping Technique',
Photogrammetric Engineering, 37 (1971), 475-80

41 Stewardson, P.B., K. Kraus and D.C. Gsell. 'DACS -
Digital Automatic Contouring System', Paper
presented to the 12th International Congress for
Photogrammetry, Ottawa, Canada, 1972

42 Collins, S.H. 'Stereoscopic Orthophoto Maps',
The Canadian Surveyor, 22 (1968), 167-76. See
also Blachut, T.J. 'Methods and Instruments
for Production and Processing of Orthophotos',
Paper presented to Commission II, 12th International
Congress for Photogrammetry, Ottawa, Canada, 1972

43 Visser, *et al.* *op cit*, 10-14

44 Hobrough, G.L. 'Automation in Photogrammetric
    Instruments', *Photogrammetric Engineering*, 31
    (1965), 595-603

45 Löscher, W. 'Wild-Raytheon Stereomat A2000 Fully
    Automatic Plotter', *Wild Reporter*, No 2 (1969), 3-5

46 Hardy, J.W. 'Automatic Stereoplotting with the
    EC-5/Planimat', *Bildmessung und Luftbildwesen*,
    No 1 (1970), 62-8

47 Brucklacher, W.A. 'Automated Control of Ortho-
    projector by Planimat with Correlator', *Bildmessung
    und Luftbildwesen*, No 2 (1968), 117-20

48 Schermerhorn, W. 'Evaluation of Present Situation
    and Prediction of Future Developments in
    Photogrammetry', *Proceedings of the ITC Post-
    Congress Seminar* (Delft, 1969), 187-91

49 Lorenz, G.G. 'Automation in Photogrammetry',
    International Archives of Photogrammetry, Series
    17 (1969)

50 Hobrough, G.L. and T.B. 'Images Correlator Speed
    Limits', *Photogrammetric Engineering*, 37 (1971),
    1045-53

51 Helava, U.V., and W.E. Chapelle. 'Epipolar-Scan
    Correlation', Paper presented to Commission II,
    12th International Congress for Photogrammetry,
    Ottawa, Canada, 1972

52 Masry, S.E. 'Digital Correlation Principles',
    Photogrammetric Engineering, 40 (1974), 303-8

53 See Colwell, R.N. (ed). *Manual of Photographic
    Interpretation*, American Society of Photogrammetry
    (1960); also Lueder, D.R. *Aerial Photographic
    Interpretation, Principles and Applications*
    (1969), 6

54 Tait, D.A. 'Photo-Interpretation and Topographic

Mapping', *Photogrammetric Record*, 6 (1970), 466-79

55 Konecny, G. 'Automation in Photogrammetry' in
    Angus-Leppan, P.V. (ed), *Control for Mapping by
    Geodetic and Photogrammetric Methods*, Report on
    Colloquium held at the University of New South
    Wales, Department of Surveying (1967), 96-119

56 Vink, A.P.A. *Some Thoughts on Photo-Interpretation*,
    ITC Publication B25 (Delft, 1964)

57 Duda, R.O. 'Elements of Pattern Recognition', in
    Mendel, J.M., and K.S. Fu (eds), *Adaptive, Learning
    and Pattern Recognition Systems: Theory and
    Applications* (1970), 3-33

58 Uhr, L. (ed). *Pattern Recognition: Theory,
    Experiment, Computer Simulations and Dynamic
    Models of Form Perception and Discovery* (1966), 1-6

59 Rosenfeld, A. 'Automated Picture Interpretation',
    in Stewart, G.A. (ed), *Land Evaluation* (1968), 187-
    99

60 Ross, D.S. 'Image-tone Enhancement', *Technical
    Papers from the 35th Annual Meeting of the American
    Society of Photogrammetry* (1969), 301-19

61 Rosen, C.A. 'Pattern Classification by Adaptive
    Machines', *Science*, 156 (1967), 38-44

62 Rosenfeld, A., *op cit*, 192-8

63 Fu, K.S. *Sequential Methods in Pattern Recognition
    and Machine Learning* (1968)

64 Rosenfeld, A. 'An Approach to Automatic Photo-
    graphic Interpretation', *Photogrammetric
    Engineering*, 28 (1962), 660-5

65 Hawkins, J.K., and C.J. Munsey. 'Automatic Photo
    Reading', *Photogrammetric Engineering*, 29 (1963),
    632-40

66 Colwell, R.N. 'The Extraction of Data from Aerial

Photographs by Human and Mechanical Means',
Photogrammetria, 20 (1965), 211-18

67   Murray, A.E.   'Perception Applications in Photo
     Interpretation', *Photogrammetric Engineering*,
     27 (1961), 627-37

68   Rib, H.T., and R.D. Miles.   'Automatic Inter-
     pretation of Terrain Features', *Photogrammetric
     Engineering*, 35 (1969), 153-64

69   Chevallier, R., A. Fontanel, G. Grau and M. Guy.
     'Application of Optical Filtering to the Study of
     Aerial Photographs', *Photogrammetria*, 26 (1970),
     17-35.   See also Nyberg, S., T. Orhaug and
     H. Svensson.   'Optical Processing for Pattern
     Properties', *Photogrammetric Engineering*, 37 (1971),
     547-59

70   Hardy, J.W., H.R. Johnston, and J.M. Godfrey.
     'An Electronic Correlator for the Planimat',
     International Archives of Photogrammetry, Series 17
     (1969).   See also Hardy, J.W.   'An Optical
     Stereoscope with Automatic Image Registration',
     ibid

71   Rosen.   *op cit*

72   Fu.   *op cit*

73   Di Pentima, A.F.   'Automation in Photo Inter-
     pretation', *Photogrammetria*, 26 (1970), 167-82

74   Bullock, F.W.   'The Photogrammetry of Bubble
     Chamber Tracks', *Photogrammetric Record*, 7 (1971),
     119-34

75   Di Pentima.   *op cit*, 180

76   Steiner, D., and H. Maurer.   'Toward a Quantitative
     Semi-Automatic System for the Photo-Interpretation
     of Terrain Cover Types', *Transactions of the Second
     International Symposium on Photo Interpretation*
     (Paris, 1966), VI.3 - VI.10

318

77 Steiner, D., and H. Maurer. 'The Use of Stereo
Height as a Discriminating Variable for Crop
Classification on Aerial Photographs',
Photogrammetria, 24 (1969), 223-41

78 Steiner, D., K. Baumberger and H. Maurer.
'Computer-Processing and Classification on
Multivariate Information from Remote Sensing
Imagery - a Review of the Methodology as Applied
to a Sample of Agricultural Crops', *Proceedings
of the 6th International Symposium on Remote
Sensing of Environment*, Vol 2 (1969), 895-907.
See also Steiner, D. 'Using the Time Dimension
for Automated Crop Surveys from Space',
*Technical Papers from the 35th Annual Meeting of
the American Society of Photogrammetry* (1969),
286-300

79 Schwartz, D.E., and F. Caspall. 'The Use of Radar
in the Discrimination and Identification of
Agricultural Land Use', *Proceedings of the 5th
Symposium on Remote Sensing of Environment*,
University of Michigan, Ann Arbor (1968), 233-47

80 Lo, C.P. 'A Typological Classification of
Buildings in the City Centre of Glasgow from
Aerial Photographs', *Photogrammetria*, 27 (1971),
135-57

81 Johnston, R.J. 'Choice in Classification: the
Subjectivity of Objective Methods', *Annals of the
Association of American Geographers*, 58 (1968),
575-89

82 Steiner, D. 'Automation in Photo Interpretation',
Geoforum, 2 (1970), 75-88

83 Lo, C.P. 'The Use of Orthophotographs in Urban
Planning', *Town Planning Review*, 44 (1973), 71-87

CHAPTER VII

1 De Hass, W.G.L. *Integrated Surveys and the Social
Sciences* (Delft, 1966)

2 Stewart, G.A. (ed). *Land Evaluation*, Macmillan of
   Australia (1968), 1

3 Mabbutt, J.A. 'Review of Concepts of Land
   Classification', in Stewart (ed), *op cit*, 11-28

4 Komarov, V.B. 'Aerial Photography in the
   Investigation of Natural Resources in the USSR',
   in Stewart, *op cit*, 143-85

5 Webster, R., and P.H.T. Beckett. 'Terrain
   Classification and Evaluation Using Air Photo-
   graphy: A Review of Recent Work at Oxford',
   Photogrammetria, 26 (1970), 51-75

6 MacPhail, D.C. 'Photomorphic Mapping in Chile',
   Photogrammetric Engineering, 37 (1971), 1139-48

7 Verstappen, H. Th. 'Geomorphology and Environment' (Inaugural Address, Delft, 1968), 15

8 Rey, P. 'Photographie aerienne et Vegetation',
   in Stewart, *op cit*, 187-207

9 Hall, A.D., and R.E. Fagan. 'Definition of System',
   in Buckley, W. (ed). *Modern Systems Research
   for the Behavioral Scientist: A Source Book*
   (Chicago, 1968), 81-92

10 Jerie, H.G. *From Photogrammetry to Photogrammetric
   System Engineering*, Inaugural Lecture, International
   Institute for Aerial Survey and Earth Sciences
   (Delft, 1969)

11 Thomas, E.N., and J.L. Shofer. *Strategies for the
   Evaluation of Alternative Transportation Plans*,
   National Cooperative Highway Research Program
   Report No 96 (1970)

12 Dueker, K.J. *Spatial Data Systems*, Technical
   Reports Nos 4-6, Urban and Transportation Informa-
   tion System, Department of Geography, Northwestern
   University, Evanston (1966)

13 Dacey, M.F., and D.F. Marble. *Some Comments on*

Certain Technical Aspects of Geographic Information Systems, Technical Report No 2, Geographic Information Systems, Department of Geography, Northwestern University (no date), 3

14 Garrison, W.L., *et al*. *Data Systems Requirements for Geographic Research*, Technical Report No 1, Geographic Information Systems, Department of Geography, Northwestern University (no date), 142

15 Moore, E.G., and B.S. Wellar. 'Urban Data Collection by Airborne Sensor', *Journal of the American Institute of Planners*, 35 (1969), 35-43

16 Jeffers, J.N.R. 'The Use of Electronic Computers in Land-Use Surveys based on Photo-Interpretation', *Photogrammetric Record*, 5 (1967), 465-9

17 Miller, C.L., and R.A. Laflamme. 'The Digital Terrain Model - Theory and Application', *Photogrammetric Engineering*, 24 (1958), 433-42

18 Colner, B.J. 'Aerial Photography in the Unified Information System', *Highway Research Record*, 142 (1966), 16-27

19 Dueker, K.J., and F.E. Horton. 'Urban-Change Detection Systems: Remote-Sensing Inputs', *Photogrammetria*, 28 (1972), 89-106

20 Wellar, B.S. 'Remote Sensing and Urban Information Systems', *Photogrammetric Engineering*, 39 (1973), 1041-50

21 Lo, C.P. 'The Use of Orthophotographs in Urban Planning', *Town Planning Review*, 44 (1973), 71-87

22 Collins, S.H. 'Stereoscopic Orthophoto Maps', *The Canadian Surveyor*, 22 (1968), 167-76; Blachut, T.J. 'Stereo-Orthophoto System', *Bildmessung und Luftbildwesen*, No 1 (1971), 25-8

23 Tomlinson, R.F. 'A Geographic Information System for Regional Planning', in Stewart, G.A. (ed). *Land Evaluation*, Macmillan of Australia (1968), 200-10

ACKNOWLEDGEMENTS

The support given by the Department of Geography and
Geology, University of Hong Kong, has been most valuable.
I am particularly grateful to Miss Betty Chun, Secretary
in the Department, for having undertaken the arduous task
of typing the whole manuscript, Messrs H.K. Kwan, M. Chiu
and T.B. Wong of the Cartographic Unit for preparing all
the illustrations, and Mr Y.S. Cheung of the Photographic
Unit for carrying out all the photographic work.
Encouragement and help have never failed to come from my
teachers and friends in the Department of Geography,
University of Glasgow, notably Messrs G. Petrie and
D.A. Tait, to whom my thanks are also due.

The author also wishes to record thanks to the
following persons, organisations and firms for generously
supplying information and photographs as well as granting
permission to use them in this book: Professors J.B. Bird
and J.T. Parry, Department of Geography, McGill University,
Montreal, Canada; Professor T.J. Blachut, Photogrammetric
Research Section, National Research Council, Canada;
Messrs Crouch and Hogg, Glasgow; Mr R.J. Hackman, United
States Geological Survey, Denver, Colorado, USA;
Mr P.G. Mott, Hunting Surveys and Consultants Ltd,
Boreham Wood, England; Allied Engineering Co, Kansas,
USA; A.S. Kongsberg Vapenfabrikk, Norway; Bausch and
Lomb, USA; BKS Surveys Ltd, Northern Ireland; Carl Zeiss
Oberkochen, West Germany; Cartographic Engineering Ltd,
England; Crown Lands and Survey Office, PWD, Hong Kong;
D-Mac Ltd, Glasgow, Scotland; Department of Geography,
University of Glasgow; Department of Geography and
Geology, University of Hong Kong; Hunting Surveys Ltd,
England; International Institute for Aerial Survey and
Earth Sciences (ITC), Enschede, the Netherlands; Itek
Corporation, USA; Kern and Co Ltd, Aarau, Switzerland;
Officine Galileo, Florence; Ottico Meccanica Italiana,

322

Rome; VEB Carl Zeiss Jena, East Germany; Wild Heerbrugg
Ltd, Switzerland.

INDEX

characteristic curve, 25, *25*

colour film, 26;
applied to agriculture and forestry, 196-8; applied to geology and geomorphology, 198-201; applied to soil surveys, 197-8; requirements in photography, 192-6

conceptual model, 15

Continuous Strip Camera, 37-8, 130

control points, 65-7; *65*, 80;
ground control, 65-7; minor control, 67-8

correlative photo-interpretation, 166

correlator, 247, 249

crab, 34, *35*

crop type identification, 153-4

cross ratio, *see* anharmonic ratio

crown closure, 120

crown diameter, 120

decision making, 255-7, *257*; adaptive network approach, 260; fixed network, 260

densitometer, 153, 199

densitometric data, 102

density, 24

determination of current velocity, 110-1

diaphragm, 20

diapositive, 32

differential rectification, 232, 238

Digital Terrain Model (DTM), 273-5

digitising, 219-27; cartographic, 225-7, 226; mode of, 225; photogram-

metric, 219-27

digitiser, 220-27;
grid type, 223; linear type, 223, *224*; rotary type, 220, *221*, *222*

dip angle, 95-7;
apparent dip, 96-7, *97*; true dip, 96-7, *97*

dipping platen, 95

directional data, 89-92

discriminant function analysis, 256-7, *257*, 261-2

d-Mac Pencil Follower, 225-6, *226*

dot grid, 124

DP-1 Double Projector, Zeiss Oberkochen, 75

drainage density, 87-8, 102

drift, 34, *35*

drop-lines, 237-9, *238*

Earth Resources Technology Satellite (ERTS), 218

electromagnetic spectrum, 178-80

epipolar scan correlation, 251-2, *252*

false colour film, 201-9, *202*;
advantages of, 203-5; applied to forestry and agriculture, 205-6; applied to land use studies, 206-7; applied to urban analysis, 208-9

farming types, 153-4

feature extraction, 255

fiducial marks, 42, *43*

film flattening, 22

film types, 24-31;
colour, 11, 178, 180, 188-201, *190*;
infra-red, 11, 180, 181-

topographic plotters, 81
townscapes, 162-3
transect method, 123
transport studies, 162
tree height measurement,
    117-20;
    shadow length method, 117
Trimetrogon photography, 17

Universal Automatic Map
Compilation System
(UNAMACE), 250
universal machines, 81
urban analysis, 171-6;
    changes, 163-5; land use,
    159-61; poverty, 175-6;
    social structure, 172-5
Urban Information System,
    275-8, *276*

vegetation survey and
mapping, 154-7
vertical photography, 16
visible spectrum, 179-80,
    179

want of correspondence, *see*
    Y-parallax

X-parallax, 57-60, *59*, 110-
    1, 115, 119, 259

Y-parallax, 56-7, *57*, *58*,
    59, 110-1, 119, 259

Zoom Transfer Scope, Bausch
and Lomb, 71-2

DATE DUE

ÉCHÉANCE